DIGITAL
ELECTRONICS

An Introduction
to Theory and Practice

DIGITAL ELECTRONICS

An Introduction to Theory and Practice

Second Edition

William H. Gothmann, P.E.

Senior Engineer
ISC Systems Corporation

PRENTICE-HALL, INC., Englewood Cliffs, N.J.   07632

Library of Congress Cataloging in Publication Data

GOTHMANN, WILLIAM H.
    Digital electronics.

    Includes index.
    1. Digital electronics.   I. Title.
TK7868.D5G67   1982   621.3815   81-12129
ISBN 0-13-212159-X   AACR2

Editorial/production supervision
        and interior design by Karen Skrable
Manufacturing buyer: Gordon Osbourne
Cover design by Jorge Hernandez

Printed in the United States of America

10   9   8   7   6   5   4   3

ISBN 0-13-212159-X

Prentice-Hall International, Inc., *London*
Prentice-Hall of Australia Pty. Limited, *Sydney*
Prentice-Hall of Canada, Ltd., *Toronto*
Prentice-Hall of India Private Limited, *New Delhi*
Prentice-Hall of Japan, Inc., *Tokyo*
Prentice-Hall of Southeast Asia Pte. Ltd., *Singapore*
Whitehall Books Limited, *Wellington, New Zealand*

**To
Myrna**

Wisdom is better than strength

—Solomon

CONTENTS

9 MEMORY CIRCUITS AND SYSTEMS *278*

10 THE DIGITAL COMPUTER *300*

11 ANALOG/DIGITAL CONVERSION *333*

USEFUL REFERENCE DATA

PREFACE

Digital electronics is increasingly encroaching on areas previously monopolized by analog techniques. Engineers and technicians, in order to meet this challenge of change, will find this book an excellent introduction to this exciting, dynamic field. It is designed to provide a thorough grounding in basic design techniques and is therefore suitable for use as a first text in digital electronics in both engineering and technical schools.

There are a number of books on the market that explain digital theory in depth. There are also a number that explain logic hardware in depth. However, theory itself is of limited value, as is a study of just the logic hardware in use today. The designer needs to know both. This book was written to bridge the gap between theory and practice in order to provide the tools required for the designer to solve digital problems today and to prepare him to solve the changing digital problems of tomorrow.

Chapter 1 provides a comparison of digital and analog techniques in instrumentation, communications, control systems, and computing. A short history of digital electronics and computers is included to give perspective to today's rapid developments.

Chapter 2 introduces the binary, octal, and hexadecimal number systems—those used by our friendly computers.

Chapter 3 introduces binary codes. Not only are numeric codes such as 8421, Gray, BCD and Hamming studied, but all the commonly used alpha-numeric ones, such as ASCII, EBCDIC, and Hollerith.

Chapter 4 introduces Boolean algebra and MIL STD-806C logic symbols. In order to provide the reader with practice interpreting truth tables and logic symbols, the method of perfect induction is used to prove most of the theorems. Additionally, since those in electronics are a curious lot, each time a logic symbol is introduced, transistor, relay, and diode circuits are shown that could be used to accomplish the logic function.

Chapter 5 leads the reader through the switching characteristics of transistors (both bipolar and field effect devices) and the logic families in use today, comparing their advantages and disadvantages. Particular attention is given to TTL and its subfamilies.

Chapter 6 introduces the reader to combinational logic and the reduction of Boolean expressions. Sum of product, product of sums, and hybrid functions are all investigated. Four algorithms are used for these reductions: level mapping, variable mapping, algebraic reduction, and the Quine-McCluskey method.

Chapter 7 opens the field of sequential logic to the reader, providing the theory and procedures for designing counters using type *T*, *D*, and *JK* flip-flops. Ripple, nonsequential, and cycle counters are all discussed in detail. A section on asynchronous logic design is introduced using procedures consistent with those used in synchronous design, providing maximum transfer of learning.

A few years ago, a digital text could end with Chapter 7. However, with the introduction of more and more MSI and LSI, understanding a flip-flop, gate, and inverter is not sufficient. Chapter 8 provides learning in medium scale functions: arithmetic units, multiplexers, decoders, and other functions. These are, after all, the chips used with LSI in designing any system.

In reviewing the text in preparation for this *Second Edition*, the increasing importance of bus structures and memory systems was quite apparent. Chapter 9 was written to meet this need, both at the IC and the subsystem level.

Chapter 10, The Digital Computer, has been extensively revised to provide more depth in microcomputers. The Motorola MC6801 is used as the subject of this discussion because of its flexibility and straight-forward instruction set.

Chapter 11 presents the principles of analog-to-digital and digital-to-analog conversion. The major methods of conversion are presented along with their specifications.

This edition of DIGITAL ELECTRONICS could not have been completed except for the valuable advice of instructors too numerous to mention from many parts of the world. I am greatly indebted to them for their constructive suggestions. Finally, I wish to thank my wife, Myrna, for her active support and encouragement in revising this text.

BILL GOTHMANN

DIGITAL
ELECTRONICS

An Introduction
to Theory and Practice

DEVELOPMENT OF DIGITAL ELECTRONICS

1

1-1 THE EARLY YEARS

The digital way of doing things has been with us ever since man discovered he has more than one finger or toe. Using his fingers (digits), he found he was able to relate the concept "two" to an appropriate number of fingers, opening a whole new world to him. He or she could, for example, relate the number of clubs needed to the number of spouses he or she wanted to woo. It was only much later that he found need for expressing parts of the whole: half of a mammoth or one-quarter of a Saber-toothed tiger. With this development, two separate mathematical systems were realized: analog and digital. This book is devoted to explaining the *digital* method of design and analysis.

1-2 DIGITAL AND ANALOG METHODS

To understand digital methods, one must first understand what makes them differ from analog methods. The term *digital* refers to any process that is accomplished using discrete units. Fingers, toes, rocks, elephants, and germs are examples. Each of these could be used as a unit or group of units to express a whole number. Addition could be performed by adding these fingers, toes, rocks, elephants, or germs.

In contrast with digital numbers, analog numbers are represented as directly measurable quantities such as volts, resistances, rotations, and distances. Note the similarity of the words *analog* and *analogy*. Both indicate a parallel between a real thing and a representation of the real thing. Thus, in the analog method the number 15 could be represented as a 150-degree rotation of the needle on a meter, the depth of water in a sinking boat, or the appropriate length on a wooden scale. Addition could be performed by adding needle rotations, water depths, or lengths of the scale.

Instrumentation

With meters, the analog method has been used extensively in electronics to represent such phenomena as intensity, frequency, speed, and time. These scale-type meters have the advantage of allowing the observer to determine the percent of full scale at a glance but have the disadvantage of requiring interpretations on the part of the observer. Not only does this require more time, but it also introduces "eyeball error" into the reading; everyone gets a different message. These problems can be overcome by using a digital readout device (Fig. 1-1). This advantage, coupled with the greatly reduced cost of integrated circuits, explains the wide use of digital voltmeters, digital panel meters, and other digital readout devices in recent years. People prefer a meter that "tells it like it is."

Communications

Today, both digital and analog methods are used in communications. In the digital method (Fig. 1-2), an analog waveform is sampled, and the digital representation of the amplitude is transmitted at time intervals sufficient to define the waveform. Multiple channels of information can be provided by alternately sampling waveforms A and B (Fig. 1-3), converting the sample amplitude to a digital number (digitizing), and then transmitting the resultant digital information. This system is called time division multiplexing (TDM), because the time is shared between channels A and B.

TDM contrasts with the straight analog method [Fig. 1-3(b)], which uses modulators that provide different carrier frequencies for each channel. These channels can then be combined in a frequency mixer and the resultant transmitted. This method of transmission is called frequency division multiplexing (FDM). Both TDM and FDM are being used today for transmission of voice information.

Control Systems

Control systems also use both digital and analog techniques. In digital systems (Fig. 1-4), digital information is picked off the motor shaft indicating its position. This is compared with the position commanded by the controller and an error signal developed at the output of the comparator indicating the difference between where the shaft is and where it should be. This error signal, in the form of digital pulses, then commands the motor to move in the proper direction to minimize the error. Similarly, in analog control systems an error signal is generated by comparing the pick-off voltage with the controller voltage. This analog error signal then commands the motor to move in the proper direction to minimize the error.

(a) Analog Volt-Ohm-Meter

(b) Digital Volt-Ohm-Meter

FIG. 1-1 Digital and Analog Meters

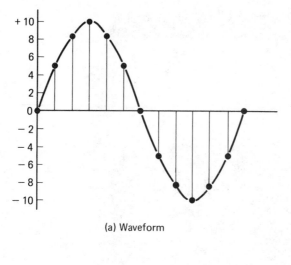

(a) Waveform

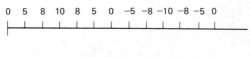

(b) Numbers Transmitted   **FIG. 1-2** Digitized Waveform

Computers

Both methods, digital and analog, are used for computing (Fig. 1-5). Digital computers, using arithmetic similar to that which can be done with paper and pencil, add, subtract, multiply, and divide, providing an output accurate to the digit. Analog computers, using voltages and potentiometer rotations to represent numbers, add, subtract, and even integrate to provide a resultant voltage. This result can then be read (and interpreted) by the observer on a panel meter or chart recorder.

1-3 HISTORY OF DIGITAL COMPUTING

Since the term *digits* refers to a discrete counting unit, it is not surprising that most of its early application was in computing devices. In the following paragraphs, we shall examine some of the history of computing science.

The Abacus

In early history, man computed by making marks in the dust—one mark for a 1, two marks for a 2, etc. The word *abacus* originally meant "dust," referring to this practice of making marks in the dust. From this early beginning. the Greeks and the

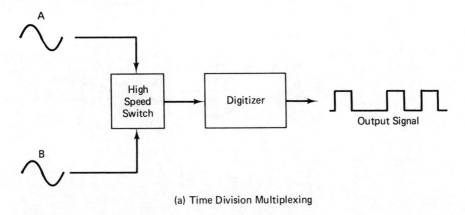

(a) Time Division Multiplexing

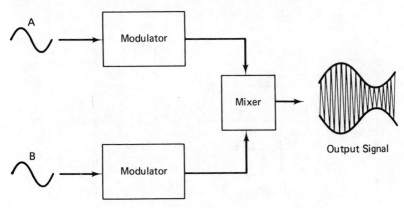

(b) Frequency Division Multiplexing

FIG. 1-3 Multiplexing Communications Channels

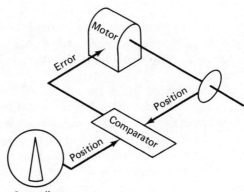

FIG. 1-4 Control System

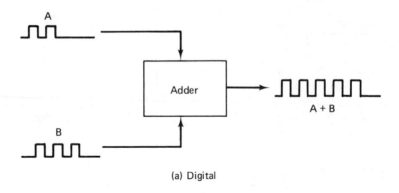

(a) Digital

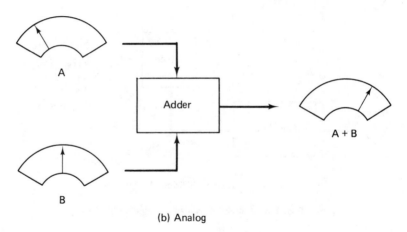

(b) Analog

FIG. 1-5 Computer Arithmetic

Romans developed a system of counting involving beads on a marble table. The beads were grouped for units, tens, hundreds, and so on. Later on, grooves were included in the marble top so the beads could be freely moved. The word *countertop* originated from this practice.

The earliest recorded use of the abacus as we know it today occurred in the fifth century, B.C. It was used extensively in both Europe and Asia until the twelfth century, when Arabic numbers became widespread. With this introduction, the abacus fell into disuse in Europe. So complete was the takeover by Arabic numerals in Europe that an officer in Napoleon's army reported "discovering" the abacus in Russia.

In countries using non-Arabic numbers the abacus is still extensively used. It is estimated that 95% of Japanese business is transacted using this remarkable instrument.

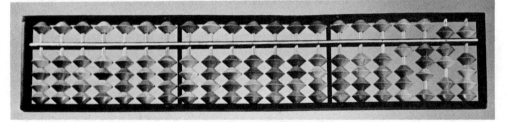

(a) Japanese Model

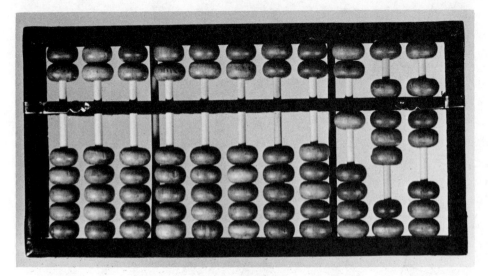

(b) Chinese Model

FIG. 1-6 The Abacus

The abacus is rather a simple digital device based upon Roman numerals. Beads above the bar [Fig. 1-6(a)] have a value of 5 when moved toward the bar. Those below the bar have a value of 1 when moved toward the bar. Thus, each column is capable of counting up to 9. The columns are arranged in the same manner as a decimal number: the right column is the units column, the next one the tens column, and so forth.

There are two types of abacuses used today: the Chinese version and the Japanese version. The Chinese version is similar to the Japanese version except that it has two 5-beads and five 1-beads in each column.

Computer Development

In 1614, John Napier, a distinguished Scottish mathematician, discovered logarithms, allowing complex mathematical functions to be calculated with ease. Logs are basically analog in nature; that is, they are an analog of the actual number.

Three years later, this same gentleman, John Napier, discovered a digital method for performing multiplication. He constructed a device using "rods" or "bones," having four sides with numbers arranged in the form shown in Fig. 1-7. To use these devices, bones are selected and arranged to display the first operand at the top of the bones. In the example shown in Fig. 1-8, the number 642 is to be multiplied by 72. The 6, 4, and 2 bones are then selected and placed beside the 1 bone as shown.

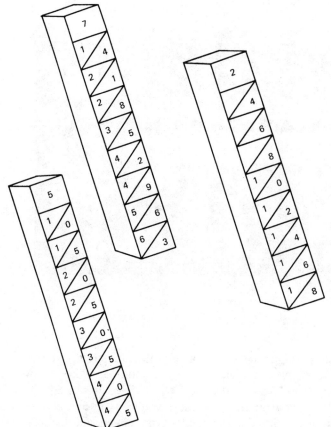

FIG. 1-7 Napier's Bones

First, the 642 is multiplied by 7. To do this, move to the 7 row on the 1 bone as shown and read the answer, 4494. Note that carries are added automatically to the next digit. Having determined the answer, multiply by 10 (since we are multiplying by 70, not 7), resulting 44940. Next, multiply by 2 using the same procedure. Add these two results together for the final answer. Although this device made history, it was only one link in a chain of developments in computing devices.

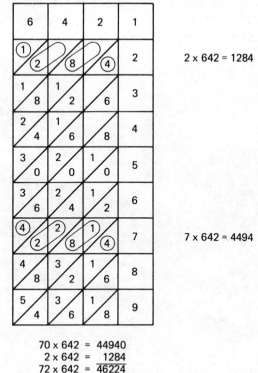

FIG. 1-8 Multiplying with Napier's
Bones

```
70 x 642  =  44940
 2 x 642  =   1284
72 x 642  =  46224
```

Fifteen years later, in 1632, William Oughtred, a noted English mathematician, invented the slide rule, also an analog device, using John Napier's logarithms as its fundamental principle.

In 1642, Blaise Pascal, a French scientist and theological author, became exasperated with adding long columns of figures for his father, who was a tax collector. Only 19 years old, he invented a desk calculator for addition (Fig. 1-9). The machine consisted of 8 gears, each having 10 teeth. But the most interesting feature of the machine was that it was capable of carrying a digit to the next higher position when a gear progressed to the zero location.

Later that century, in 1670, Baron Gottfried Wilhelm von Leibniz, a German mathematician, building on Pascal's invention, invented a machine he called a stepped reckoner, capable of adding, subtracting, multiplying, dividing, and extracting roots. This development was the dawn of the mechanical calculators used in the twentieth century.

In 1780, Frenchman Joseph Marie Jacquard began building on a 1725 idea of Basile Bouchon, developing an automatic loom capable of weaving intricate designs into fabric as programmed by punched cards. In 1801 he exibited a perfected machine at the industrial exposition in Paris, capable of controlling as many as 12,000 needles at once. This automatic loom remains relatively unchanged to this day. The punched cards he used are still a primary method of input to modern-day computers, a hole being a 1 and no hole being a 0.

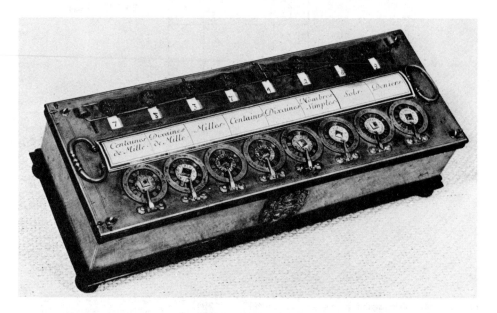

FIG. 1-9 Pascal's Calculator (Courtesy International Business Machines Corporation)

One of the most important nineteenth-century figures in computer science was Charles Babbage, a British mathematician. Between 1812 and 1822 he invented a machine he called a "difference engine" for the purpose of calculating mathematical tables. He actually built a small model of the machine. In 1813 he obtained the backing of the Royal Society for building a much larger machine with seven 20-digit registers (places to store numbers) and a printed output. Because of manufacturing difficulties, the Royal Society withdrew its support from the project in 1822.

Undaunted, he continued working, developing an improved concept, his "analytical engine," in 1833 (Fig. 1-10). A remarkably advanced device, this universal calculator was to contain a 1000-location memory, each location capable of storing up to 50 digits in counting wheels. It was to use Jacquard's punched card concept for control. Branching (skipping forward or backward in the stack of punched cards) could be accomplished by testing the numerical sign of the result. A dial would indicate how many cards to skip.

The arithmetic unit was to be capable of adding or subtracting in 1 second or performing a 50 by 50-digit multiplication in 1 minute. Babbage spent years developing a simultaneous carry capability.

Numbers were to be fed into the unit by punched cards and manual setting of

FIG. 1-10 Babbage's Analytical Engine (Courtesy British Crown Copyright. Science Museum, London)

dials. Output was to be in the form of punched cards, printed copy, or stereotype molds.

Babbage died in 1871, having built only a few parts of his machine because his mechanical tolerances could not be met by the machine shops of that day. However, the concept—program, storage, arithmetic unit, input, and output—forms the basis for all modern digital computers.

In 1876, English scientist William Thompson, also known as Lord Kelvin, developed the first analog computer for solving differential equations. Although his design was sound, he could not produce a working model due to insufficient development of technology at that time. In 1930, Vannevar Bush of the Massachusetts Institute of Technology produced an analog computer based upon Thompson's design.

In 1887, Herman Hollerith was a statistician working for the Bureau of Census in Buffalo, New York. Since the census was to be taken again in 3 years, he developed a system whereby census information was to be entered onto punched cards. These cards would then be inserted into a card reader consisting of needles that tested for holes in the card (Fig. 1-11). When a hole existed, the needle projected through the card, contacting a pool of mercury below. This would, in turn, cause a dial to advance one position on the display panel. Hollerith's device was used in the 1890 census, reducing the time required for analysis to one-third that of the 1880 census even

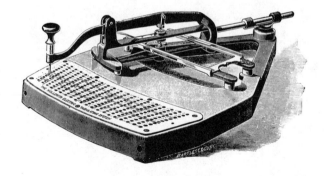

(a) Hollerith Pantagraph Punch

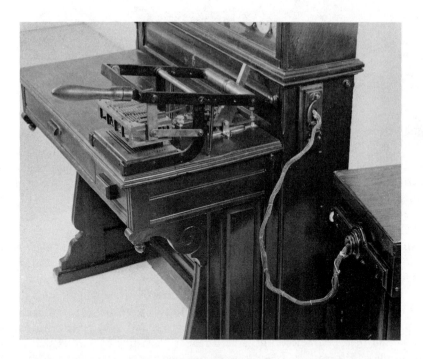

(b) Card Reader

FIG. 1-11 Hollerith's Card Processors (Courtesy International Business Machines Corporation)

though the population had grown by 26 %. Later, Hollerith developed a manufacturing company that was, ultimately, to become part of International Business Machines (IBM). The coding used for punched cards is still referred to as Hollerith code.

In 1937, Howard H. Aiken developed a "computer" composed of 78 interconnected adding machines controlled by a "piano roll" type of punched program. It used standard calculator parts, relays, and mechanical counters; it had a 300-millisecond timing cycle, used a 4-horsepower drive, and weighed 5 tons. Called the Mark I, it was used by the U.S. Army until the end of World War II.

In 1946, the first vacuum-tude digital computer, the Electronic Numerical Integrator And Computer (ENIAC), was developed by J. Presper Eckert and John W. Mauchly at the Moore School of Engineering, University of Pennsylvania (Fig. 1-12).

FIG. 1-12 Electronic Numerical Integrator and Computer (ENIAC) (Courtesy Sperry Corporation)

It contained 18,000 tubes and was capable of 5000 additions or 500 multiplications per second.

In 1948, IBM built the Selective Sequence Electronic Calculator (SSEC). Containing 12,500 vacuum tubes, it was the first computer to employ a stored program.

In 1951, the first commercial data processing computer was built, the Sperry Rand UNIVAC I, containing semiconductor diodes and vacuum tubes (Fig. 1-13). It was used by the Bureau of the Census for processing the 1950 census. By 1955, over 15 UNIVAC I's had been produced. Meanwhile, IBM had started commercially producing computers, and we were ushered into the computer age.

FIG. 1-13 UNIVAC I Computer (Courtesy Sperry Corporation)

In the early 1970's the development of the microprocessor revolutionized the computer industry. Because of that event and the subsequent development of the microcomputer, our lives will never be the same.

Today, hundreds of types of computers are being produced, ranging in size from pocket-sized calculators to room-sized multiprocessor installations (Figs. 1-14, 1-15, and 1-16) and ranging from tens of dollars to millions of dollars.

FIG. 1-14 HP-65 Programmable Pocket Calculator (Courtesy Hewlett-Packard)

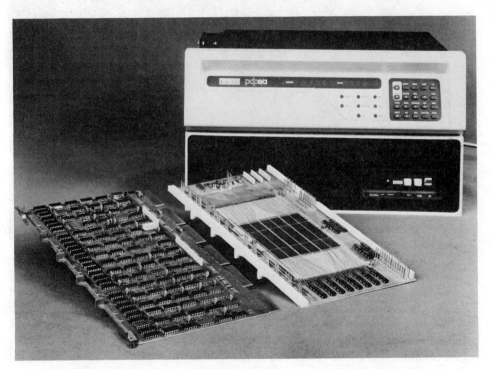

FIG. 1-15 Digital Equipment Corporation PDP-8/A Minicomputer (Courtesy Digital Equipment Corporation)

FIG. 1-16 UNIVAC 1110 Data Processing Facility (Courtesy Sperry Corporation)

1-4 SUMMARY

The digital method of arithmetic originated with man using his fingers (digits) as counting devices. This system differs from the analog method, which uses directly measurable quantities to express numbers. Both systems are used in instrumentation, communications, control systems, and computing systems. The abacus was an early digital calculator. John Napier, Blaise Pascal, and Baron von Leibniz contributed greatly to expanding the knowledge of computing science. However, it was left to Charles Babbage to set down the principles for the digital computer and to Herman Hollerith to put data processing techniques to use. Since then, computers have developed from the Mark I, ENIAC, and SSEC to modern high-speed machines capable of performing millions of operations every second.

1-5 PROBLEMS

1-1. State one advantage of an analog meter. A disadvantage.

1-2. State one advantage of a digital meter. A disadvantage.

1-3. Define the term *analog*.

1-4. Define the term *digital*.

1-5. Why are many phases of electronics converting to digital processes?

1-6. What did the term *abacus* originally mean?

1-7. What number is on the abacus shown in Fig. 1-6(a)?

1-8. Match the following men with their inventions:

A. Charles Babbage	1. Adding machine
B. Herman Hollerith	2. Analog computer
C. Joseph Jacquard	3. Calculating bones or rods
D. Baron von Leibniz	4. Card reader for census
E. John Napier	5. Difference engine
F. William Oughtred	6. Multiplying/dividing machine
G. Blaise Pascal	7. Programmable loom
H. William Thompson	8. Slide rule

NUMBER SYSTEMS

2

2-1 NUMBER SYSTEMS USED IN DIGITAL ELECTRONICS

In Chapter 1 we introduced digital techniques, including what digital is, where it is used, and who developed it for computers. The very term *digital* implies a system of counting using discrete units. Therefore, in this chapter we shall introduce four systems of arithmetic that are used in digital systems:

A. Decimal
B. Binary
C. Hexadecimal
D. Octal

Since digital systems use the binary system extensively, in this chapter we shall examine both fixed-point and floating-point binary arithmetic.

2-2 THE DECIMAL NUMBER SYSTEM

Before examining other number systems in detail, the reader should observe several important characteristics of the decimal system. We have become so accustomed to its use that its orderliness can easily be overlooked. The decimal number system contains 10 unique symbols: 0, 1, 2, 3, 4, 5, 6, 7, 8, and 9. To many readers, this is not an earth-shattering revelation. Note that, although we call it a decimal (10's) system, it does not have a symbol for "ten" (10) but expresses it and any number above 10 as a combination of these symbols. Because it has 10 symbols, its radix (or base) is said to be 10.

Even though the decimal system has only 10 symbols, any number of any magnitude can be expressed by using our system of positional weighting. Consider the number 3472. It can be broken down as follows:

$$3472 = 3000 \quad + 400 \quad + 70 \quad + 2$$
$$= 3 \times 10^3 + 4 \times 10^2 + 7 \times 10^1 + 2 \times 10^0$$

Note that because the 3 is positioned four places to the left of the decimal point, it has a much greater weight than the 2. If the 3 were next to the point, it would be worth only 3 instead of 3000. Thus, the position of the digit with reference to the decimal point determines its weight. This can be illustrated further by rearranging the digits of the previous example; the value of the number changes:

$$4237 = 4000 \quad + 200 \quad + 30 \quad + 7$$
$$= 4 \times 10^3 + 2 \times 10^2 + 3 \times 10^1 + 7 \times 10^0$$

The principle of positional weighting can be extended to any number system. Any number can be represented by the equation

$$Y = d_n \times r^n + d_{n-1} \times r^{n-1} + \ldots + d_1 \times r^1 + d_0 \times r^0 \qquad (2\text{-}1)$$

where Y is the value of the entire number, d_n is the value of the nth digit from the point, and r is the radix, or base. This equation has already been applied to the base 10 number system. Next, we shall apply it to the binary (base 2) system.

2-3 THE BINARY NUMBER SYSTEM

Assume the number 9 must be transmitted from the town of Sasquatch to the town of Spartania. There are several ways the number could be electrically transmitted:

A. A 9 volt (V) signal could be sent and then measured at Spartania. This would be an analog method.
B. A digital method using 10 lines could be employed. Line 9 would be energized, turning on lamp 9 at the distant end (Fig. 2-1). This system has distinctive advantages over the analog system. Noise, line resistance, and meter errors become irrelevant. All that is important is whether the light is on or off. Whether the lamp is bright or dim does not matter. However, it has the disadvantage of requiring 100 lines for 100 unique numbers.
C. Combinations of lines could be used, minimizing the number of lines required (Fig. 2-2). In this manner, 16 unique numbers could be sent using only 4 lines. This system overcomes the disadvantage of B above.

The last system of transmission makes maximum use of the lines available. All combinations of four lines are used, and the system appears very orderly. Next, we shall convert the X's to ones (1's) and the blanks to zeros (0's) such that a 1 indi-

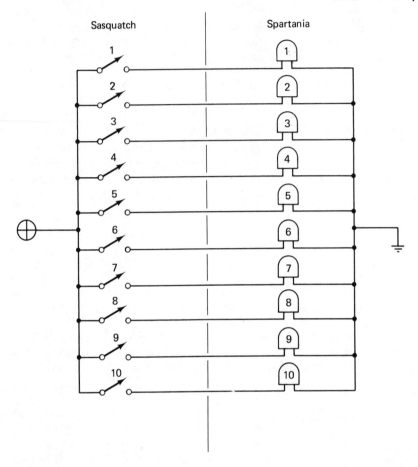

FIG. 2-1 Digital Transmission Using Discrete Lines

cates an energized line and a 0 an unenergized line (Table 2-1). This system of numbering is called a binary system, because each position (line) can take on only one of
two values: 0 or 1. These positions are called *bits*, a contraction of the words *binary
digits*. Look closely at this binary counting sequence. It is very similar to the decimal
system of counting. Starting at 0, the next number is 1. The right-hand bit has now
assumed the highest value within the system, 1. The next number in the sequence will
require that a 1 be added to the column-2 bit and that the column-1 bit be set to 0.
This is similar to the decimal system when counting from 9 to 10. In this case a 1 is
added to the 10's column and the units column set to 0. This binary sequence continues until the decimal number 15 (1111) is reached, the highest number that can be
represented by four bits.

The primary advantage of using this binary counting system as opposed to the
10-discrete-line method is that it minimizes the number of lines required. This is
also a property of the decimal number system. In decimal, a straight counting se-

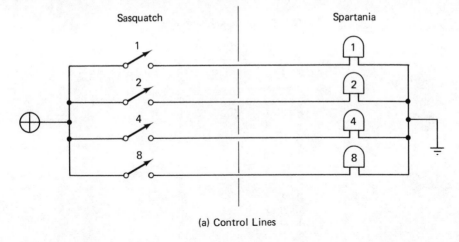

(a) Control Lines

8	4	2	1	Number
				0
			X	1
		X		2
		X	X	3
	X			4
	X		X	5
	X	X		6
	X	X	X	7
X				8
X			X	9
X		X		10
X		X	X	11
X	X			12
X	X		X	13
X	X	X		14
X	X	X	X	15

(b) Numbers Transmitted

FIG. 2-2 Digital Transmission Using Combinations

quence can be used in order to express 100 different numbers using digits 0 through 9. Thus, sequential counting assures that a minimum number of lines or digits will be used, regardless of the base of the system.

Binary numbers are used extensively throughout all digital systems, because of the very nature of electronics. A 1 can be represented by a saturated transistor, a light turned on, a relay energized, or a magnet magnetized in a particular direction. A 0 can be represented as a cutoff transistor, a light turned off, a de-energized relay, or the magnet magnetized in the opposite direction. In each of these, there are only two values that the device can assume. For this reason, we shall study the binary system of numbers in detail.

TABLE 2-1. Binary Numbers

Binary				Decimal
8	4	2	1	
0	0	0	0	0
0	0	0	1	1
0	0	1	0	2
0	0	1	1	3
0	1	0	0	4
0	1	0	1	5
0	1	1	0	6
0	1	1	1	7
1	0	0	0	8
1	0	0	1	9
1	0	1	0	10
1	0	1	1	11
1	1	0	0	12
1	1	0	1	13
1	1	1	0	14
1	1	1	1	15

Converting Binary to Decimal

Since humans have 10 fingers and use the decimal system, there is a communications gap between a binary electronic system and *Homo sapiens*. This means that those of us who deal in 1's and 0's must be capable both of converting our orderly binary system into the decimal system and of converting decimal numbers into our binary system.

Like the decimal system, the binary system is positionally weighted. Each position represents a particular value of 2^n. Table 2-2 illustrates both binary and decimal positional weighting. The base is indicated by the subscripts. Because of this property

TABLE 2-2. Binary and Decimal Positional Weighting

(a) Binary	(b) Decimal
$1_2 = 1 \times 2^0 = 1_{10}$	$1_{10} = 1 \times 10^0 = 1_{10}$
$10_2 = 1 \times 2^1 = 2_{10}$	$10_{10} = 1 \times 10^1 = 10_{10}$
$100_2 = 1 \times 2^2 = 4_{10}$	$100_{10} = 1 \times 10^2 = 100_{10}$
$1000_2 = 1 \times 2^3 = 8_{10}$	$1000_{10} = 1 \times 10^3 = 1000_{10}$
$10000_2 = 1 \times 2^4 = 16_{10}$	$10000_{10} = 1 \times 10^4 = 10000_{10}$

of positional weighting, the procedure for converting a binary number to decimal is very similar to that of breaking a decimal number into its weighted values, as discussed in Sec. 2-2. It merely requires substitution of numbers into Eq. (2-1). The d's will all be 0's or 1's, the r's will all be 2 (the radix), and the n's will be various powers of 2, depending on the position of the digit with reference to the binary point. (In binary it is referred to as a binary point, not a decimal point.)

Example 2-1: Convert 10111_2 to decimal.

SOLUTION:

$$Y = d_4 \times r^4 + d_3 \times r^3 + d_2 \times r^2 + d_1 \times r^1 + d_0 \times r^0$$
$$= 1 \times 2^4 + 0 \times 2^3 + 1 \times 2^2 + 1 \times 2^1 + 1 \times 2^0$$
$$= 16 + 0 + 4 + 2 + 1$$

$$10111_2 = 23_{10}$$

Example 2-2: Convert 1011101001_2 to decimal.

SOLUTION:

$$Y = d_9 \times r^9 + d_8 \times r^8 + d_7 \times r^7 + d_6 \times r^6 + d_5 \times r^5 +$$
$$d_4 \times r^4 + d_3 \times r^3 + d_2 \times r^2 + d_1 \times r^1 + d_0 \times r^0$$
$$= 1 \times 2^9 + 0 \times 2^8 + 1 \times 2^7 + 1 \times 2^6 + 1 \times 2^5 +$$
$$0 \times 2^4 + 1 \times 2^3 + 0 \times 2^2 + 0 \times 2^1 + 1 \times 2^0$$
$$= 512 + 0 + 128 + 64 + 32 + 0 + 8 + 0 + 0 + 1$$

$$1011101001_2 = 745_{10}$$

The above method will always provide the correct decimal representation of a binary number. There is a second method, called the *dibble-dobble method*, that will also provide the solution. To use this method, start with the left-hand bit. Multiply this value by 2, and add the next bit to the right. Multiply this value by 2, and add the next bit to the right. Stop when the binary point is reached.

Example 2-3: Convert 110111_2 to decimal.

SOLUTION:

Copy down the left bit:	1
Multiply by 2, add next bit	$(2 \times 1) + 1 = 3$
Multiply by 2, add next bit	$(2 \times 3) + 0 = 6$
Multiply by 2, add next bit	$(2 \times 6) + 1 = 13$
Multiply by 2, add next bit	$(2 \times 13) + 1 = 27$
Multiply by 2, add next bit	$(2 \times 27) + 1 = 55$

Therefore, $110111_2 = 55_{10}$.

The above two methods provide the means for converting binary numbers to decimal numbers. However, in digital circuits, it is necessary to convert from decimal to binary also.

Converting Decimal to Binary

There are two procedures for converting from decimal to binary. These are the reverse of the two procedures used to convert binary to decimal.

The first method requires a table of powers of 2 (Table 2-3). Because of this

TABLE 2-3. Powers of 2

2^n	n	2^{-n}
1	0	1.0
2	1	0.5
4	2	0.25
8	3	0.125
16	4	0.062 5
32	5	0.031 25
64	6	0.015 625
128	7	0.007 812 5
256	8	0.003 906 25
512	9	0.001 953 125
1 024	10	0.000 976 562 5
2 048	11	0.000 488 281 25
4 096	12	0.000 244 140 625
8 192	13	0.000 122 070 312 5
16 384	14	0.000 061 035 156 25
32 768	15	0.000 030 517 578 125
65 536	16	0.000 015 258 789 062 5
131 072	17	0.000 007 629 394 531 25
262 144	18	0.000 003 814 697 265 625
524 288	19	0.000 001 907 348 632 812 5
1 048 576	20	0.000 000 953 674 316 406 25
2 097 152	21	0.000 000 476 837 158 203 125
4 194 304	22	0.000 000 238 418 579 101 562 5
8 388 608	23	0.000 000 119 209 289 550 781 25
16 777 216	24	0.000 000 059 604 644 775 390 625
33 554 432	25	0.000 000 029 802 322 387 695 312 5
67 108 864	26	0.000 000 014 901 161 193 847 656 25
134 217 728	27	0.000 000 007 450 580 596 923 828 125
268 435 456	28	0.000 000 003 725 290 298 461 914 062 5
536 870 912	29	0.000 000 001 862 645 149 230 957 031 25
1 073 741 824	30	0.000 000 000 931 322 574 615 478 515 625
2 147 483 648	31	0.000 000 000 465 661 287 307 739 257 812 5
4 294 967 296	32	0.000 000 000 232 830 643 653 869 628 906 25
8 589 934 592	33	0.000 000 000 116 415 321 826 934 814 453 125
17 179 869 184	34	0.000 000 000 058 207 660 913 467 407 226 562 5
34 359 738 368	35	0.000 000 000 029 103 830 456 733 703 613 281 25
68 719 476 736	36	0.000 000 000 014 551 915 228 366 851 806 640 625

restriction, it is more useful for small numbers where these powers have been memorized. Starting with the decimal number to be evaluated, obtain the largest power of 2 from the table without exceeding the original number. Record this. Then subtract this table-obtained number from the original number. Repeat the process for the remainder, and continue until the remainder is zero. Finally, add the binary numbers obtained from the table. The result is the answer.

Example 2-4: Convert 43_{10} to binary.

SOLUTION:

From Table 2-3, 32 is the largest $32 =$ 100000
number without exceeding 43.

$$43 - 32 = 11$$

From the table, 8 is the largest $8 = \quad 1000$
number without exceeding 11.

$$11 - 8 = 3$$

From the table, 2 is the largest $2 = \quad\quad 10$
number without exceeding 3.

$$3 - 2 = 1$$

One is the largest number. $\underline{1 = \quad\quad\quad 1}$
Add the binary numbers. $43 = \quad 101011_2$

Example 2-5: Convert 200_{10} to binary.

SOLUTION:

$$
\begin{aligned}
& 128 = 10000000 \\
200 - 128 = 72 \qquad & 64 = 1000000 \\
72 - 64 = 8 \qquad & \underline{8 = \quad\ 1000} \\
& 200 = 11001000
\end{aligned}
$$

Therefore, $200_{10} = 11001000_2$.

The second method of converting decimal to binary is the opposite of the dibble-dobble method. In using this method, the number is successively divided by 2 and its remainders recorded. The final binary result is obtained by assembling all the remainders, with the last remainder being the most significant bit (MSB).

Example 2-6: Convert 43_{10} to binary.

SOLUTION:

Successive Division	Remainders
2)43	
2)21	1
2)10	1
2) 5	0
2) 2	1
2) 1	0
0	1

Reading the remainders from the bottom to the top,

$$43_{10} = 101011_2$$

This is the same result as in Example 2-4.

Example 2-7: Convert 200_{10} to binary.

SOLUTION:

Successive Division	Remainders
2)200	
2)100	0
2) 50	0
2) 25	0
2) 12	1
2) 6	0
2) 3	0
2) 1	1
0	1

Reading the remainders from the bottom to the top, the result is

$$200_{10} = 11001000_2$$

This is the same result as in Example 2-5.

Binary Addition and Subtraction

In decimal addition, numbers are added according to an addition table. In a similar manner, the binary system uses the addition table shown in Table 2-4. Addition is accomplished in a manner similar to that in decimal.

TABLE 2-4. Binary Addition Table

	0	1
0	0	1
1	1	10

Example 2-8: Add 1011_2 and 110_2.

SOLUTION:

In the 1's column, $1 + 0 = 1$
In the 2's column, $1 + 1 = 0$
with carry of 1
In the 4's column, $1 + 0 + 1 = 0$
with a carry of 1
In the 8's column, $1 + 1 = 10$
Therefore, $1011_2 + 110_2 = 10001_2$.

Column Numbers

```
  8 4 2 1

  1 0 1 1
+   1 1 0
1 0 0 0 1
```

Example 2-9: Add 11110_2 and 11_2.

SOLUTION:

```
  11110
+    11
 100001
```

Binary subtraction is performed in a manner similar to that in decimal subtraction. Because the binary system has only two digits, binary subtraction requires more borrowing operations than decimal.

Example 2-10: Subtract one from 100_2.

SOLUTION:

$$Column\ Numbers$$
$$4\quad 2\quad 1$$

$$\not{1}\quad \overset{1}{\not{0}}\quad {}^1 0$$
$$\underline{\qquad\quad 1}$$
$$1\quad 1$$

Since, in the 1's column, a 1 cannot be subtracted from a 0, a 1 must be borrowed from the 2's column. But it is also a 0, so a 1 must be borrowed from the 4's column, making the 4's column 0 and the 2's column 10. A 1 can now be borrowed from the 2's column, making it a 1 and the 1's column 10.

$$10 - 1 = 1 \text{ in the 1's column.}$$

$$1 - 0 = 1 \text{ in the 2's column.}$$

$$0 - 0 = 0 \text{ in the 4's column.}$$

Therefore, $100_2 - 1_2 = 11_2$.

Binary Multiplication

There are two methods for binary multiplication: the *paper* method and the *computer* method. Multiplication in either method requires use of the multiplication table shown in Table 2-5. The paper method is a method similar to that which one would use when multiplying two decimal numbers on paper.

TABLE 2-5. Binary Multiplication Table

	0	1
0	0	0
1	0	1

Example 2-11: Multiply 1011_2 by 101_2.

SOLUTION:

$$
\begin{array}{r}
1\,0\,1\,1 \\
\times\ \ 1\,0\,1 \\
\hline
1\,0\,1\,1 \\
0\,0\,0\,0 \\
1\,0\,1\,1 \\
\hline
1\,1\,0\,1\,1\,1
\end{array}
$$

Example 2-12: Multiply 11010_2 by 11011_2.

SOLUTION:

$$
\begin{array}{r}
1\ 1\ 0\ 1\ 0 \\
\times 1\ 1\ 0\ 1\ 1 \\
\hline
1\ 1\ 0\ 1\ 0 \\
1\ 1\ 0\ 1\ 0 \\
0\ 0\ 0\ 0\ 0 \\
1\ 1\ 0\ 1\ 0 \\
1\ 1\ 0\ 1\ 0 \\
\hline
1\ 0\ 1\ 0\ 1\ 1\ 1\ 1\ 1\ 0
\end{array}
$$

This paper method cannot be used by a digital computer, for a computer can add only two numbers at a time (plus carries). In addition, the computer has only a fixed number of bit positions available. Assume the number 1011_2 must be multiplied by 1010_2. Multiplying a four-bit number by another four-bit number yields an eight-bit result. Thus, a four-bit multiplicand register (an electronic device capable of holding these four bits) and an eight-bit multiplier/result register are required. Referring to Table 2-6, the multiplier is placed in the left four bits of the MQ (multiplier/quotient) register, and the right four bits are set to zero. This allows the MQ to be used as a dual-use register: it holds the multiplier and the partial (then final) results. The multiplicand is placed in the M register. The MQ register is then shifted one bit to the left. If a 1 is shifted out, M is then added to the MQ register. If a 0 is shifted out, a 0 is added to the MQ register. The entire process is repeated four times (the number of bits in the multiplier). The result appears in the MQ register.

TABLE 2-6. Computer Binary Multiplication

MQ Register	1010	0000
Shift MQ	1 0100	0000
Add M		1011
Sum in MQ	0100	1011
Shift MQ	0 1001	0110
Add 0		0000
Sum in MQ	1001	0110
Shift MQ	1 0010	1100
Add M		1011
Sum in MQ	0011	0111
Shift MQ	0 0110	1110
Add 0		0000
Sum in MQ	0110	1110

Binary Division

Like multiplication, division can be accomplished by two methods: a paper method and a computer method. In the paper method, long-division procedures similar to those in decimal are used.

Example 2-13: Divide 110110_2 by 101_2.

SOLUTION:

$$
\begin{array}{r}
1010 \\
101)\overline{110110} \\
\underline{101} \\
111 \\
\underline{101} \\
\overline{100} \quad \text{remainder}
\end{array}
$$

The computer method of division requires successive subtraction. Assume an eight-bit dividend is to be divided by a four-bit divisor. The quotient will be formed in the right half of the MQ register and the remainder in the left half. The dividend is first placed in the MQ register and the divisor in the D register (Table 2-7). The divisor is then subtracted from the dividend. The result is considered positive if the most significant bit (the far left bit) is a 0 and negative if it is a 1. If this result is positive, then an error has occurred, for the quotient would be greater than four bits.

TABLE 2-7. Computer Division

MQ	0010 1011	
Subtract D	0110	
MQ	1100 1011	Result is negative; the
Add D	0110	division is valid
MQ	0010 1011	
Shift MQ left	0101 0110	
Subtract D	0110	
MQ	1111 0110	Result is negative
Add D	0110	
MQ	0101 0110	Put a zero in quotient
Shift MQ left	1010 1100	
Subtract D	0110	
MQ	0100 1100	Result is positive
Add 1 to quotient	1	
MQ	0100 1101	
Shift MQ left	1001 1010	
Subtract D	0110	
MQ	0011 1010	Result is positive
Add 1	1	
MQ	0011 1011	
Shift MQ left	0111 0110	
Subtract D	0110	
MQ	0001 0110	
Add 1	1	
MQ	0001 0111	Final answer
	Remainder Quotient	

If the result is negative, then the quotient will be four or less bits, sufficiently small to be contained in a four-bit register.

The MQ register is next shifted left one bit and the divisor subtracted from it. If the result is positive, a 1 is added to the least significant bit (LSB) of the MQ register where the quotient is accumulated. If it is negative, the divisor is added to the MQ and the MQ shifted left one bit. This effectively puts a 0 in this bit of the quotient. The process is continued until the MQ register has been shifted left four bits (the number of bits in the D register). The remainder is then in the left half of the MQ register and the quotient in the right half.

2-4 THE HEXADECIMAL NUMBER SYSTEM

Binary numbers are long—not opinion, but fact. To represent the number 4096_{10}, 12 bits must be used. These numbers are fine for stupid machines, but for humans, they are simply too bulky. Imagine that a computer is programmed with all the social security numbers of the people in the United States and that, in order to get your social security check, you must ask the operator to feed your number in binary into the machine. One such number is

$$1,1111,1111,0111,1010,1000,0110,1001$$

Hexadecimal numbers are one-fourth this length and yet provide the same binary information, allowing easy people-to-people communication.

The hexadecimal number system was born out of the need to express binary numbers concisely and is by far the most commonly used number system in computer literature. The hexadecimal number is formed from a binary number (word) by grouping bits in groups of four bits each, starting at the binary point. This is a logical way of grouping, since computer words come in 8 bits, 12 bits, 16 bits, 32 bits, and so on. In a group of 4 bits, the decimal numbers 0 — 15 can be represented (Table 2-8). However, each combination must be represented by a unique symbol. So, drawing on American genius, individuality, and bravery, the symbols A, B, C, D, E, and F were selected. Don't you think the hexadecimal number 1F,F7A,869 would be easier to give to a computer operator than the above 29-bit binary number?

Converting Binary to Hexadecimal

Binary numbers can be easily converted to hexadecimal by grouping in groups of four starting at the binary point.

Example 2-14: Convert 1010111010_2 to hexadecimal.

SOLUTION:

Group in fours	10,1011,1010
Convert each number	2 B A
Thus, the solution is 2BA.	

Table 2-8. Hexadecimal Number System

Binary	Hexadecimal	Decimal
0000	0	0
0001	1	1
0010	2	2
0011	3	3
0100	4	4
0101	5	5
0110	6	6
0111	7	7
1000	8	8
1001	9	9
1010	A	10
1011	B	11
1100	C	12
1101	D	13
1110	E	14
1111	F	15
1 0000	10	16

Example 2-15: Convert 11011110101110_2 to hexadecimal.

SOLUTION:

Group in fours	11,0111,1010,1110
Convert each number	3 7 A E

Thus, the solution is 37AE.

Converting Hexadecimal to Binary

Hexadecimal numbers can be converted to binary by converting each digit.

Example 2-16: Convert $4A8C_{16}$ to binary.

SOLUTION:

Given:	4 A 8 C
Convert each digit:	0100 1010 1000 1100

Thus, the solution is $100,1010,1000,1100_2$.

Example 2-17: Convert $FACE_{16}$ to binary.

SOLUTION:

Given:	F A C E
Convert each digit:	1111 1010 1100 1110

Thus, the solution is $1111,1010,1100,1110_2$.

Converting Hexadecimal to Decimal

The hexadecimal number system is, like its neighbors decimal and binary, a positionally weighted system obeying Eq. (2-1), where the radix is 16. To convert from hexadecimal (usually called hex for short) to decimal, merely substitute the digits into the equation.

Example 2-18: Convert $2C9_{16}$ to decimal.

SOLUTION:

$$2C9_{16} = d_2 \times r^2 + d_1 \times r^1 + d_0 \times r^0$$
$$= 2 \times 16^2 + 12 \times 16^1 + 9 \times 16^0$$
$$= 512 + 192 + 9$$
$$= 713_{10}$$

Note that a C represents a decimal 12, as indicated in Table 2-8.

Example 2-19: Convert $EB4A_{16}$ to decimal.

SOLUTION:

$$EB4A_{16} = d_3 \times r^3 + d_2 \times r^2 + d_1 \times r^1 + d_0 \times r^0$$
$$= 14 \times 16^3 + 11 \times 16^2 + 4 \times 16^1 + 10 \times 16^0$$
$$= 57,344 + 2816 + 64 + 10$$
$$= 60,234_{10}$$

Converting Decimal to Hexadecimal

Any decimal number can be converted to hex by successively dividing by 16. The remainders can then be converted to hex and read up from the bottom to obtain the hexadecimal results.

Example 2-20: Convert 423_{10} to hexadecimal.

SOLUTION:

Successive Division	Remainders	Hex Notation
16)423		
16) 26	7	7
16) 1	10	A
0	1	1

Reading the remainders up from the bottom, the result is $1A7_{16}$.

Example 2-21: Convert 72905_{10} to hexadecimal.

SOLUTION:

Successive Division	Remainders	Hex Notation
16)72905		
16) 4556	9	9
16) 284	12	C
16) 17	12	C
16) 1	1	1
0	1	1

Reading the remainders up from the bottom, the result is $11,CC9_{16}$.

Hexadecimal Addition and Subtraction

The noble hex is added by referring to a hex addition table (Table 2-9). Addition is then performed in a manner similar to that in decimal addition.

Example 2-22: Add $1A8_{16}$ and $67B_{16}$.

SOLUTION:

Column A:

$$8 + B = 8 + 11_{10}$$
$$= 19_{10}$$
$$= 16 + 3$$
$$= 13_{16}$$

Sum of 3, carry of 1.

Columns

C	B	A
1	A	8
+ 6	7	B
8	2	3

Column B:

$$1 + A + 7 = 1 + 10_{10} + 7$$
$$= 18_{10}$$
$$= 16 + 2$$
$$= 12_{16}$$

Sum of 2, carry of 1.

Column C:

$$1 + 1 + 6 = 8$$

Sum of 8, no carry.

Thus, $1A8_{16} + 67B_{16} = 823_{16}$.

TABLE 2-9. Hexadecimal Addition Table

	0	1	2	3	4	5	6	7	8	9	A	B	C	D	E	F
0	0	1	2	3	4	5	6	7	8	9	A	B	C	D	E	F
1	1	2	3	4	5	6	7	8	9	A	B	C	D	E	F	10
2	2	3	4	5	6	7	8	9	A	B	C	D	E	F	10	11
3	3	4	5	6	7	8	9	A	B	C	D	E	F	10	11	12
4	4	5	6	7	8	9	A	B	C	D	E	F	10	11	12	13
5	5	6	7	8	9	A	B	C	D	E	F	10	11	12	13	14
6	6	7	8	9	A	B	C	D	E	F	10	11	12	13	14	15
7	7	8	9	A	B	C	D	E	F	10	11	12	13	14	15	16
8	8	9	A	B	C	D	E	F	10	11	12	13	14	15	16	17
9	9	A	B	C	D	E	F	10	11	12	13	14	15	16	17	18
A	A	B	C	D	E	F	10	11	12	13	14	15	16	17	18	19
B	B	C	D	E	F	10	11	12	13	14	15	16	17	18	19	1A
C	C	D	E	F	10	11	12	13	14	15	16	17	18	19	1A	1B
D	D	E	F	10	11	12	13	14	15	16	17	18	19	1A	1B	1C
E	E	F	10	11	12	13	14	15	16	17	18	19	1A	1B	1C	1D
F	F	10	11	12	13	14	15	16	17	18	19	1A	1B	1C	1D	1E

Example 2-23: Add $ACEF1_{16}$ and $16B7D_{16}$.

SOLUTION:

$$
\begin{array}{r}
ACEF1 \\
+\ 16B7D \\
\hline
C3A6E
\end{array}
$$

Subtraction in hexadecimal also requires reference to the addition table. To use the table, find the smaller of the two numbers along the left edge. Then move horizontally along the row until the higher of the two numbers is found. The result is found at the top of this row. For example, find B — 8. Find the 8 along the left edge. Move along the row until the B is found. The answer, 3, appears at the top of the column. Now apply this to a subtraction problem.

Example 2-24: Subtract 3A8 from 1273.

SOLUTION:

$$
\begin{array}{r}
1273 \\
-\ 3A8 \\
\hline
ECB
\end{array}
$$

Hexadecimal Multiplication

Did you ever hear of a hex multiplying? To do it requires reference to the hex multiplication table (Table 2-10).

TABLE 2-10. Hexadecimal Multiplication Table

	0	1	2	3	4	5	6	7	8	9	A	B	C	D	E	F
0	0	0	0	0	0	0	0	0	0	0	0	0	0	0	0	0
1	0	1	2	3	4	5	6	7	8	9	A	B	C	D	E	F
2	0	2	4	6	8	A	C	E	10	12	14	16	18	1A	1C	1E
3	0	3	6	9	C	F	12	15	18	1B	1E	21	24	27	2A	2D
4	0	4	8	C	10	14	18	1C	20	24	28	2C	30	34	38	3C
5	0	5	A	F	14	19	1E	23	28	2D	32	37	3C	41	46	4B
6	0	6	C	12	18	1E	24	2A	30	36	3C	42	48	4E	54	5A
7	0	7	E	15	1C	23	2A	31	38	3F	46	4D	54	5B	62	69
8	0	8	10	18	20	28	30	38	40	48	50	58	60	68	70	78
9	0	9	12	1B	24	2D	36	3F	48	51	5A	63	6C	75	7E	87
A	0	A	14	1E	28	32	3C	46	50	5A	64	6E	78	82	8C	96
B	0	B	16	21	2C	37	42	4D	58	63	6E	79	84	8F	9A	A5
C	0	C	18	24	30	3C	48	54	60	6C	78	84	90	9C	A8	B4
D	0	D	1A	27	34	41	4E	5B	68	75	82	8F	9C	A9	B6	C3
E	0	E	1C	2A	38	46	54	62	70	7E	8C	9A	A8	B6	C4	D2
F	0	F	1E	2D	3C	4B	5A	69	78	87	96	A5	B4	C3	D2	E1

Example 2-25: Multiply $1A3_{16}$ by 89_{16}.

SOLUTION:

$$
\begin{array}{r}
1A3 \\
\times\ 89 \\
\hline
EBB \\
D18 \\
\hline
E03B
\end{array}
$$

Hexadecimal Division

Hexadecimal division is accomplished by using decimal methods and the hex multiplication table (Table 2-10).

Example 2-26: Divide $1EC87_{16}$ by $A5_{16}$.

SOLUTION:

$$
\begin{array}{r}
2FC \\
A5\overline{)1EC87} \\
14A \\
\hline
A28 \\
9AB \\
\hline
7D7 \\
7BC \\
\hline
1B \quad \text{remainder}
\end{array}
$$

2-5 THE OCTAL NUMBER SYSTEM

The octal number system was used extensively by early minicomputers. However, in both large and small systems, it has largely been supplanted by the hexadecimal system. The octal system is formed by grouping bits in groups of 3, starting at the binary point. Table 2-11 compares the binary, decimal, hexadecimal, and octal systems. Note that the highest digit within the octal system is 7 and there are no 8's or 9's.

TABLE 2-11. Number Systems

Binary	Decimal	Hexadecimal	Octal
0	0	0	0
1	1	1	1
10	2	2	2
11	3	3	3
100	4	4	4
101	5	5	5
110	6	6	6
111	7	7	7
1000	8	8	10
1001	9	9	11
1010	10	A	12
1011	11	B	13
1100	12	C	14
1101	13	D	15
1110	14	E	16
1111	15	F	17
1 0000	16	10	20

Binary-Octal Conversions

Conversion from binary to octal proceeds from the foregoing definitions. Simply divide the number into groups of three bits each, starting at the binary point. Then express each group by its decimal (or octal) equivalent. Note that the highest digit in the octal system is a 7. (Do you remember that the highest digit in the decimal system is a 9?)

Example 2-27: Convert 11111011110101_2 to octal.

SOLUTION:

Divide into groups of three.   11,111,011,110,101
Express each group in decimal.   3   7   3   6   5
Therefore, $11111011110101_2 = 37,365_8$.

Example 2-28: Convert 1011110100011000111_2 to octal.

SOLUTION:

Divide into groups.   1,011,110,100,011,000,111
Express groups in decimal.   1   3   6   4   3   0   7
Therefore, $1011110100011000111_2 = 1,364,307_8$.

Conversion from octal to binary is the reverse of this process. Express each octal number in its appropriate binary notation.

Example 2-29: Convert 3674_8 to binary.

SOLUTION:

Copy the octal number.   3   6   7   4
Convert each to binary.   011   110   111   100
Therefore, $3674_8 = 11,110,111,100_2$.

2-6 BINARY ARITHMETIC IN COMPUTERS

Up to this point, the assumption has been made that the whole world is composed of whole, positive integers. However, how can my negative bank balance be represented in the computer if this is the case? Or how can four bits (this time I mean $0.50) be represented? The following paragraphs illustrate number systems for expressing both negative and fractional quantities in binary.

Two's Complement Arithmetic

Since both positive and negative numbers must be represented in a computer, an additional bit, called the sign bit, can be added to each number. If this bit is a 0, the number is positive. If it is a 1, the number is negative. This system is called a signed magnitude system. For example,

$$000011 = +3$$
$$100011 = -3$$
$$001010 = +10$$
$$101010 = -10$$

However, under this system a great deal of manipulation is necessary to add a positive number to a negative number. This makes the number system possible but impractical.

One of the essential characteristics of a computer is its finite word length, since, when it is initially designed, hardware is provided only for a specific number of bits with which to represent a number. This same principle can be applied to the decimal system. Assume that only two places were available. Counting up from the bottom, the numbers would be 0, 1, 2, . . . , 9, 10, 11, . . . , 97, 98, 99, 0, 1, 2, ... Note that all carries out from the 10's column would be discarded. Addition would be performed as follows:

$$
\begin{array}{r}
31 \\
+\ 33 \\
\hline
64
\end{array}
\qquad\qquad
\begin{array}{r}
64 \\
+\ 73 \\
\hline
37
\end{array}
$$

Again, note that all carries out from the 10's column are lost.

This type of math is called modulus arithmetic. Returning to the binary number system, the result of adding eight-bit words using modulus arithmetic would be as follows:

$$
\begin{array}{r}
1011\ 1011 \\
0100\ 0100 \\
\hline
1111\ 1111
\end{array}
\qquad\qquad
\begin{array}{r}
1010\ 0001 \\
1101\ 0111 \\
\hline
0111\ 1000
\end{array}
$$

As in the decimal system, carries out from the eighth bit are discarded.

The two's complement system is used to represent negative numbers using modulus arithmetic. Assume an eight-bit number length. Negative numbers are represented by the transformation $2^8 - N$, where N is the positive representation of the number. The exponent, 8, was chosen because 2^8 results in a number that is 1 larger than the eight bits can express.

Example 2-30: Express -5 in two's complement form.

SOLUTION:

$$
\begin{array}{ll}
2^8 = & 1\ 0000\ 0000 \\
\text{Subtract 5} & -\quad 0000\ 0101 \\
\hline
\text{Result} & \quad 1111\ 1011
\end{array}
$$

Thus, the two's complement of 5 is 1111 1011.

The foregoing principles can be applied to any number of any bit length. To express any number in two's complement, subtract it from 2^P, where P is the word length.

Example 2-31: Express the numbers -4, -15, and -17 in 12-bit two's complement form.

SOLUTION:

$$-4 = 1\ 0000\ 0000\ 0000 - 0100 = 1111\ 1111\ 1100$$

$$-15 = 1\ 0000\ 0000\ 0000 - 1111 = 1111\ 1111\ 0001$$

$$-17 = 1\ 0000\ 0000\ 0000 - 0001\ 0001 = 1111\ 1110\ 1111$$

Example 2-32: Express the decimal number $-16,000$ in 16-bit two's complement form.

SOLUTION:

$2^{16} =$	1 0000 0000 0000 0000
Subtract 16,000	$-$ 0011 1110 1000 0000
Answer	1100 0001 1000 0000

Besides subtracting from 2^p, there are two other methods of computing the two's complement of a number. One is to complement all the bits (change all the 1's to 0's and all the 0's to 1's) and then add 1.

Example 2-33: Express -4 in two's complement form.

SOLUTION:

Positive expression of number	0000 0100
Complement	1111 1011
Add 1	1111 1100

Therefore, -4 equals 1111 1100.

Compare this result with that in Example 2-31.

Example 2-34: Express -17 in two's complement form.

SOLUTION:

Positive expression of number	0001 0001
Complement	1110 1110
Add 1	1110 1111

Compare this result with that in Example 2-31.

The third method of converting a positive number to its two's complement is as follows:

Starting at the least significant bit and working from right to left, copy each bit down up to and including the first 1 bit encountered. Then complement the remaining bits.

Example 2-35: Express -4 in two's complement form.

SOLUTION:

Original number	0000 0100
Copy up to first 1	100
Complement remaining bits	1111 1100

Compare this result with Examples 2-31 and 2-33.

Example 2-36: Express -17 in two's complement form.

SOLUTION:

Original number	0001 0001
Copy up to first 1	1
Complement remaining bits	1110 1111

Compare this result with Examples 2-31 and 2-34.

Two's complement numbers have some very interesting characteristics. Consider a four-bit word length. The decimal equivalents would be as shown in Table 2-12.

TABLE 2-12. Two's Complement Numbers

Binary	Decimal
0111	+7
0110	+6
0101	+5
0100	+4
0011	+3
0010	+2
0001	+1
0000	0
1111	−1
1110	−2
1101	−3
1100	−4
1011	−5
1010	−6
1001	−7
1000	−8

Observe the following from the table:

A. There is one unique 0.
B. The two's complement of 0 is 0.
C. The leftmost bit cannot be used to express quantity. It is a sign bit such that if it is a 1, the number is negative, and if it is a 0, the number is positive.
D. There are eight negative integers, seven positive integers, and one 0, making a total of 16 unique states. In the two's complement system, there will always be $2^{P-1} - 1$ positive integers, 2^{P-1} negative integers, and one 0, for a total of 2^P unique states, where P is the number of bits in the binary word.
E. Significant information is contained in the 1's of the positive numbers and the 0's of the negative numbers. Think about that for a while. (*Hint:* Consider -2 and $+2$ in a 16-bit number.)
F. To convert a negative number to a positive number, find its two's complement.

Example 2-37: What negative value does 1001 1011 represent?

SOLUTION:

The two's complement of 1001 1011 is 0110 0101. This represents a 101_{10}. Therefore, $1001\ 1011_2 = -101_{10}$.

Now that you know how to find the two's complement and you know some "positively thrilling" characteristics of this system, what good is it? Its value lies in its use in arithmetic operations. Addition can be performed directly without any transformations, with the sign bit taking active part in the process. Using eight-bit two's complement, add -5 to $+5$. The result is, of course, zero.

$$
\begin{array}{ll}
+5 & 0000\ 0101 \\
-5 & \underline{1111\ 1011} \\
\text{Sum} & 0000\ 0000
\end{array}
$$

Since the carry is the ninth bit, it is discarded. This example illustrates that it is possible to treat a negative number as just another number to be added, with the sign bit taking active part in the addition.

Example 2-38: Add -17 to -30.

SOLUTION:

$$
\begin{array}{rl}
-17 = & 1110\ 1111 \\
-30 = & \underline{1110\ 0010} \\
-47 = & 1101\ 0001
\end{array}
$$

To check the result, convert it back into a positive binary number. The two's complement of 1101 0001 is 0010 1111, which is 47. Thus, 1101 0001 was indeed the correct answer.

Example 2-39: Add -20 to $+26$.

SOLUTION:

$$
\begin{array}{rl}
-20 = & 1110\ 1100 \\
+26 = & \underline{0001\ 1010} \\
+6 = & 0000\ 0110
\end{array}
$$

Example 2-40: Add -29 to $+14$.

SOLUTION:

$$
\begin{array}{rl}
-29 = & 1110\ 0011 \\
+14 = & \underline{0000\ 1110} \\
-15 = & 1111\ 0001
\end{array}
$$

Since the two's complement of 1111 0001 is 0000 1111, which is 15, 1111 0001 is a -15.

One's Complement Arithmetic

Although most computers use two's complement arithmetic, some use one's complement. The one's complement of a number is found by changing all the 0's of the word to 1's and all the 1's to 0's. This complemented value represents the negative of the original number. This system is very easy to implement in hardware by merely feeding all bits through inverters.

Example 2-41: Find the one's complement form of −13.

SOLUTION:

$$+13 = 0000\ 1101$$
$$-13 = 1111\ 0010$$

This system is called the one's complement because the number can be subtracted from 1111 1111 to obtain the result. The answer will be the same as if all the 1's were changed to 0's and all the 0's to 1's.

Example 2-42: Find the one's complement representation of −13.

SOLUTION:

$$
\begin{array}{r}
1111\ 1111 \\
\text{Subtract 13} \quad - 0000\ 1101 \\
\hline
1111\ 0010
\end{array}
$$

Note that this is the same result as in Example 2-41.

One of the difficulties of using one's complement is its representation of zero. Both 0000 0000 and its one's complement, 1111 1111, represent zero. The 0000 0000 is called *positive zero*, and the 1111 1111 is called *negative zero*.

Addition is performed in one's complement by (a) performing the addition as in a straight binary problem and (b) adding a 1 to the result if there is a carry out.

Example 2-43: Add −3 to −2 in one's complement using four bits.

SOLUTION:

$$
\begin{array}{lr}
-3 = & 1100 \\
-2 = & 1101 \\
\text{Simple sum} & \overline{1\ 1001} \\
\text{Add carry} & \underline{\quad\quad 1} \\
\text{Final sum} & 1010
\end{array}
$$

To check whether 1010 is the correct answer, find its one's complement, 0101. Since this is +5, 1010 was the correct answer, −5.

Example 2-44: Add −3 to +2 in one's complement using four bits.

SOLUTION:

$$
\begin{array}{rl}
-3 = & 1100 \\
+2 = & \underline{0010} \\
\text{Sum} & 1110
\end{array}
$$

This represents a -1 in one's complement.

Example 2-45: Add $+3$ to -2 in one's complement using eight bits.

SOLUTION:

$$
\begin{array}{rl}
+3 = & 0000\ 0011 \\
-2 = & \underline{1111\ 1101} \\
\text{Simple sum} & 1\ 0000\ 0000 \\
\text{Add carry} & \underline{\qquad\qquad 1} \\
\text{Final sum} & 0000\ 0001
\end{array}
$$

Binary Fractions

Thus far, the assumption has been made that everything comes in wholes. However, early in man's history he discovered he must somehow express one-half a mammoth or one-fourth a dinosaur. Although this can be done within an analog system, we do it all the time within the decimal system. The number 89.31 is an exact representation of a quantity, and it is represented using digits on this page. The binary system also has this capability.

In an earlier paragraph, the decimal number 3472 was represented as follows:

$$
\begin{aligned}
3472 &= 3000 + 400 + 70 + 2 \\
&= 3 \times 10^3 + 4 \times 10^2 + 7 \times 10^1 + 2 \times 10^0
\end{aligned}
$$

In the same manner, 34.72 can be represented as

$$
\begin{aligned}
34.72 &= 30 + 4 + 0.7 + 0.02 \\
&= 3 \times 10^1 + 4 \times 10^0 + 7 \times 10^{-1} + 2 \times 10^{-2}
\end{aligned}
$$

As can be seen, decimal fractions are an extension of the same rules used for decimal integers. Consequently, any number within any number system can be represented by the equation

$$
\begin{aligned}
Y = {} & d_n \times r^n + d_{n-1} \times r^{n-1} + \cdots + d_1 \times r^1 \\
& + d_0 \times r^0 + d_{-1} + r^{-1} + d_{-2} \times r^{-2} + \cdots \\
& + d_{-m+1} \times r^{-m+1} + d_{-m} \times r^{-m}
\end{aligned}
\tag{2-2}
$$

where Y is the value of the entire number, d_n is the value of the nth digit to the left of the point, r is the radix, and d_{-m} is the value of the mth digit to the right of the point. Applying this equation to the binary system, the number 101.11 can be represented by the following decimal equivalent:

$$Y = 1 \times 2^2 + 0 \times 2^1 + 1 \times 2^0 + 1 \times 2^{-1} + 1 \times 2^{-2}$$
$$= \quad 4 \quad + \quad 0 \quad + \quad 1 \quad + \quad 0.5 \quad + \quad 0.25$$
$$= 5.75_{10}$$

As can be seen, Eq. (2-2) forms the basis for converting binary integers and binary fractions to decimal.

Example 2-46: Convert 1101.000101_2 to decimal.

SOLUTION:

$$Y = 1 \times 2^3 \ + 1 \times 2^2 \ + 0 \times 2^1 \ + 1 \times 2^0 \ +$$
$$0 \times 2^{-1} + 0 \times 2^{-2} + 0 \times 2^{-3} + 1 \times 2^{-4} +$$
$$0 \times 2^{-5} + 1 \times 2^{-6}$$
$$Y = 8 + 4 + 0 + 1 + 0 + 0 + 0 + \tfrac{1}{16} + 0 + \tfrac{1}{64}$$
$$Y = 13 + 0.0625 + 0.015625$$
$$Y = 13.078125$$

The foregoing method demonstrates how to convert from the binary system to the decimal system. But what if the decimal number 0.375 must be converted into binary? The procedure is to successively multiply the decimal fraction by the radix (2 in this case) and collect all numbers to the left of the decimal point, reading down.

Example 2-47: Convert 0.375_{10} to binary.

SOLUTION:

Original number	0.375
Multiply 0.375 by 2	0.750
Multiply 0.75 by 2	1.50
Multiply 0.50 by 2	1.00

Reading the numbers to the left of the point down,

$$0.375_{10} = 0.011_2$$

Example 2-48: Convert 0.54545_{10} to binary.

SOLUTION:

Original number	0.54545
Multiply 0.54545 by 2	1.0909
Multiply 0.0909 by 2	0.1818
Multiply 0.1818 by 2	0.3636
Multiply 0.3636 by 2	0.7272
Multiply 0.7272 by 2	1.4544
Multiply 0.4544 by 2	0.9088
Multiply 0.9088 by 2	1.8176
Multiply 0.8176 by 2	1.6352
Multiply 0.6352 by 2	1.2704
Multiply 0.2704 by 2	0.5408

This particular number can never be expressed exactly in binary. Therefore, reading the numbers to the left of the point down,

$$0.54545_{10} = 0.1000101110_2 \text{ to 10 places}$$

As the previous example illustrates, many decimal fractions cannot be represented as exact binary numbers. In fact, the number must end in 5 to be so expressed.

To convert a mixed number such as 38.21 to binary, first convert the integer, then the fraction.

Example 2-49: Convert 38.21 to binary.

SOLUTION:

Convert the 38:

Successive Division	*Remainders*
2)38	
2)19	0
2) 9	1
2) 4	1
2) 2	0
2) 1	0
0	1

Reading up, the result is 100110.
Convert the 0.21:

Successive Multiplication

0.21
0.42
0.84
1.68
1.36
0.72
1.44
0.88

Reading down, the fraction is (0.0011010).
Adding the integer and the fraction,

$$\begin{array}{r} 100110.0000000 \\ 0.0011010 \\ \hline 100110.0011010 \end{array}$$

Therefore, 38.21_{10} equals 100110.0011010_2.

Binary fractions can be added, subtracted, multiplied, and divided, just as their decimal brothers.

Example 2-50: Add 11.011_2 and 10.001_2.

SOLUTION:

$$
\begin{array}{r}
11.011 \\
+\ 10.001 \\
\hline
101.100
\end{array}
$$

Example 2-51: Multiply 10.001_2 by 0.11_2.

SOLUTION:

$$
\begin{array}{r}
10.001 \\
\times\ \ \ \ 0.11 \\
\hline
10001 \\
10001 \\
\hline
1.10011
\end{array}
$$

Double Precision Numbers

In a 16-bit computer, numbers from $+32,767$ to $-32,768$ can be expressed in each register. However, some provision must be made for expressing numbers greater than this. Double precision provides the solution. In this system, two storage locations are used to represent each number. The format is usually as follows:

First word | S | High-order bits |

Second word | 0 | Low-order bits |

S is the sign bit and 0 is a zero. This allows a 31-bit number length with 16-bit registers. If this is insufficient, triple precision can be used, requiring three words for each number to be stored.

Floating-point Numbers

In decimal, very large and very small numbers are expressed in scientific notation: 3.59×10^{-23} and 7.42×10^{16} are examples. Binary numbers can also be expressed in this same notation by stating a number (mantissa) and an exponent of 2. The format for one computer is as shown in Fig. 2-3. In this machine the word consists of two parts: a 10-bit mantissa and a 6-bit exponent. The mantissa is in two's complement form; the leftmost bit can, therefore, be thought of as the sign bit. The binary point is assumed to be to the right of this sign bit. The 6 bits of the exponent could represent

Mantissa										Exponent					
0	1	0	0	1	0	0	0	0	0	1	0	0	0	1	1

FIG. 2-3 Floating-point Format

0 through 63. However, to express negative exponents, the number 32_{10} (100000_2) has been added to the desired exponent. This is a common system used in floating-point formats. It represents excess-32 notation. The following are examples of this system used for the exponent part of the computer word:

Desired Exponent	Binary Representation
-32	000000
-1	011111
0	100000
$+6$	100110
$+31$	111111

According to these definitions, the floating-point number in Fig. 2-3 is

The mantissa portion is	$+$ 0.100100000
The exponent portion is	100011
Subtracting 100000,	000011
The entire number is	

$$N = +0.1001_2 \times 2^3$$

$$= 100.1_2$$

$$= 4.5_{10}$$

Example 2-52: What floating-point number does 0100101000010101 represent?

SOLUTION:

The mantissa is	$+0.100101000$
The exponent is	010101
Subtracting 100000,	110101
The entire number is	

$$N = +0.100101_2 \times 2^{-11}$$

$$= +0.000,000,000,001,001,01_2$$

$$= +0.000,282,287,597,656,25_{10}$$

There are many formats of floating-point numbers, each computer having its own. Some use two words for the mantissa and one for the exponent; others use one and one-half words for the mantissa and one-half word for the exponent. On some machines, the programmer can select from several formats, depending on the accuracy desired. Some use excess-n notation for the exponent; some use two's complement. Some even use signed magnitude for both the mantissa and the exponent.

Floating-point numbers have the advantage of expressing very large and very small numbers. In the above-described machine, the mantissa is 10 bits long and the exponent 6 bits. The system is, therefore, capable of 9-bit accuracy (allowing 1 bit for the sign). The exponent bits add nothing to accuracy, only to magnitude. Fixed-point two's complement numbers expressed in 16 bits are accurate to 15 bits. Thus, although floating-point numbers can be much larger or smaller than an equivalent-length fixed-point number, they are less accurate.

To keep the full 10 bits of accuracy in the floating-point number, the most significant bit must be placed next to the sign bit. Under these conditions, the number is said to be normalized. Most computers normalize the result of any floating-point operation.

2-7 CONVERSION ALGORITHMS

An algorithm is a method of arriving at an answer that always works. Two basic algorithms have been presented in this chapter for converting from one number system to another. The first was the formula

$$Y = d_n \times r^n + d_{n-1} \times r^{n-1} + \cdots + d_0 \times r^0$$
$$+ d_{-1} \times r^{-1} + \cdots + d_{-m} \times r^{-m}$$

In this algorithm, the numbers from number system A are first converted to those of system B. Then calculation takes place within system B. For example, to convert from the binary system to the decimal system, each digit (a 1 or a 0) is converted to decimal (it happens that this is the same value in binary) and then multiplied by the radix (expressed in the decimal system) raised to some decimal power. The operations could be described as (a) convert, (b) arithmetic, and (c) collect the answer.

The second algorithm required successive division for integers. Using this method, one performs the arithmetic in number system A and then converts the results to system B. For example, to convert from decimal to hexadecimal, the arithmetic is performed in decimal, the remainders are converted to hex, and then the answer is collected. Figure 2-4 illustrates the procedures for both algorithms.

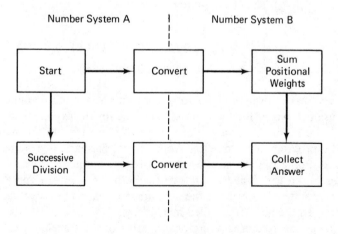

FIG. 2-4 Conversion Algorithms

Both algorithms are universal. Decimal to octal conversion is usually done using successive division, because it is easier to do the arithmetic in the decimal system. If, however, the arithmetic is done in octal, the positional weighting method can be used.

Example 2-53: Convert 395_{10} to octal using positional weights.

SOLUTION:

$$395_{10} = 300 + 90 + 5$$
$$= 3 \times 12_8^2 + 11_8 \times 12_8^1 + 5 \times 12_8^0$$
$$= 3 \times 144_8 + 11_8 \times 12_8 + 5$$
$$= 454_8 + 132_8 + 5$$
$$= 613_8$$

Conversion from octal to decimal is usually done using the positional weighting method. However, if the arithmetic is done in octal, the same answer can be found using successive division.

Example 2-54: Convert 1427_8 to decimal using successive division.

SOLUTION:

Using all octal division,

Successive Division	Octal Remainders	Decimal
12)1427		
12) 117	1	1
12) 7	11	9
0	7	7

Reading up, the result is 791_{10}. This is the same result as is obtained by summing positional weights.

2-8 SUMMARY

Digital systems require use of decimal, binary, octal, and hexadecimal number systems. The binary system is used by the hardware; the octal and hexadecimal number systems are used as a shorthand language for representing binary numbers. Each of the systems uses the two basic conversion algorithms, and each can perform all the arithmetic operations. Most computers use two's complement to represent negative numbers and binary fractions to represent fixed-point numbers less than 1. Double precision can increase the range of numbers to be expressed. However, floating-point numbers provide the widest range of binary numbers.

2-9 PROBLEMS

2-1. Convert the following binary numbers to decimal:
    (a) 1010   (b) 1110101   (c) 10101111

2-2. Convert the following binary numbers to decimal:
    (a) 1100   (b) 10010101   (c) 11011100

2-3. Convert the following decimal numbers to binary:
    (a) 23   (b) 100   (c) 145

2-4. Convert the following decimal numbers to binary:
    (a) 19   (b) 121   (c) 161

2-5. Add the following binary numbers:
    (a) 1101 + 101   (b) 1010 + 1110   (c) 11001 + 10011

2-6. Add the following binary numbers:
    (a) 101011 + 110101   (b) 111010 + 101111   (c) 111011 + 100001

2-7. Subtract the following binary numbers:
    (a) 1010 − 111   (b) 11100 − 101   (c) 101010 − 1010

2-8. Subtract the following binary numbers:
    (a) 1100 − 11   (b) 110011 − 100   (c) 111001 − 1100

2-9. Multiply the following binary numbers:
    (a) 1101 × 101   (b) 10101 × 110   (c) 110110 × 1100

2-10. Multiply the following binary numbers:
    (a) 1001 × 110   (b) 11100 × 100   (c) 101001 × 110

2-11. Divide the following binary numbers:
    (a) 11011 by 101   (b) 11100 by 100

2-12. Convert the following hexadecimal numbers to binary:
    (a) A6   (b) A07   (c) 7AB4

2-13. Convert the following hexadecimal numbers to binary:
    (a) BE   (b) BC9   (c) 9BC8

2-14. Convert the following binary numbers to hexadecimal:
    (a) 10011011101   (b) 1111011101011011   (c) 11010111010111

2-15. Convert the following binary numbers to hexadecimal:
    (a) 1010110110111   (b) 10110111011011   (c) 1111101110101111

2-16. Convert the following hexadecimal numbers to decimal:
    (a) A6   (b) A13B

2-17. Convert the following hexadecimal numbers to decimal:
    (a) E9   (b) 7CA3

2-18. Convert the following decimal numbers to hexadecimal:
    (a) 132   (b) 2352

2-19. Convert the following decimal numbers to hexadecimal:
    (a) 206   (b) 3619

2-20. Convert the following hexadecimal numbers to octal:
(a) 38AC   (b) 7FD6   (c) ABCD

2-21. Add the following hexadecimal numbers:
(a) 38 + AC   (b) 7DE1 + A786

2-22. Add the following hexadecimal numbers:
(a) 93 + DE   (b) ABCD + EF12

2-23. Subtract the following hexadecimal numbers:
(a) 1A from 39   (b) 1A9 from 7683

2-24. Subtract the following hexadecimal numbers:
(a) 2F from 46   (b) AAC from 8427

2-25. Multiply the following hexadecimal numbers:
(a) 6A × DD   (b) EC × 39

2-26. Convert the following octal numbers to binary:
(a) 123   (b) 3527

2-27. Convert the following octal numbers to binary:
(a) 7642   (b) 7015   (c) 3576

2-28. Convert the following binary numbers to octal:
(a) 111010   (b) 110110101   (c) 1101100001

2-29. Convert the following binary numbers to octal:
(a) 11001   (b) 10011101   (c) 111010111

2-30. Find the eight-bit two's complement form of the following decimal numbers:
(a) −10   (b) −52   (c) −123

2-31. Find the eight-bit two's complement form of the following decimal numbers:
(a) −14   (b) −49   (c) −99

2-32. Subtract the following decimal numbers using eight-bit two's complement arithmetic:
(a) 28 from 35   (b) 103 from 110

2-33. Subtract the following decimal numbers using eight-bit two's complement arithmetic:
(a) 35 from 66   (b) 17 from 135

2-34. Convert the following decimal numbers to 10-bit binary:
(a) 37.31   (b) 6.215   (c) 33.333

2-35. Convert the following decimal numbers to 10-bit binary:
(a) 27.2   (b) 3.21   (c) 63.362

2-36. Convert the following binary numbers to decimal:
(a) 101.01   (b) 101.1101   (c) 11101.1111

2-37. Convert the following binary numbers to decimal:
(a) 110.11   (b) 110.1001   (c) 10101.0101

2-38. Using the format shown in Fig. 2-3, convert the following floating-point numbers to binary, fixed-point numbers:
(a) 0110110000100101 (b) 1101101000011101

2-39. Using the format shown in Fig. 2-3, convert the following floating-point numbers to binary, fixed-point numbers:
(a) 0101101000101011 (b) 1100101101010111

BINARY
CODES

3

3-1 SQUARE PEGS AND ROUND HOLES

The binary system works fine for transistors, relays, switches, and integrated circuits. However, when it is to be used by 10-fingered, 10-toed people for their decimal system, it must be custom-designed to fit their system. In this chapter we shall discuss some of the methods used to express both numbers and letters as binary codes.

3-2 WEIGHTED BINARY CODES

Weighted binary codes are those which obey the positional weighting principles discussed in Sec. 2-3. Each position of the number represents a specific weight. The straight binary counting sequence is an example, for each column has a weight, 8, 4, 2, or 1.

Several systems of codes are used to express the decimal digits, 0 through 9, see Table 3-1. The 8421, 2421, and 5211 are all weighted codes, each four-bit group representing one decimal digit, the left three being weighted. The number 762_{10}, for example, would be represented in 8421 code as

$$\begin{array}{ccc} 0111 & 0110 & 0010 \\ 7 & 6 & 2 \end{array}$$

This allows any series of decimal numbers to be represented as a series of BCD codes. Using these codes, computers can add in what appears to be decimal and provide decimal answers. One application of an 8421 code appeared at a nuclear rocket test site where BCD lights indicated the time of day for observers 2 miles away. The time 6:32:40 was represented as

0000   0110   0011   0010   0100   0000
 0   6   3   2   4   0

Other codes may be more easily processed by the particular hardware involved. This could make a 2421 code or a 5211 code more attractive than the 8421 code.

TABLE 3-1. Binary-Coded Decimal Numbers

Decimal	8421	2421	5211	XS3
0	0000	0000	0000	0011
1	0001	0001	0001	0100
2	0010	0010	0011	0101
3	0011	0011	0101	0110
4	0100	0100	0111	0111
5	0101	1011	1000	1000
6	0110	1100	1010	1001
7	0111	1101	1100	1010
8	1000	1110	1110	1011
9	1001	1111	1111	1100

Reflective Codes

A code is said to be reflective when the code for 9 is the complement for the code for 0, 8 for 1, 7 for 2, 6 for 3, and 5 for 4. Note that the 2421, 5211, and XS3 codes are reflective, whereas the 8421 code is not. Reflectivity is desirable in a code when the nine's complement must be found, such as in nine's complement subtraction.

Sequential Codes

A code is said to be sequential when each succeding code is one binary number greater than its preceding code. This greatly aids mathematical manipulation of data. The 8421 and XS3 codes are sequential, whereas the 2421 and 5211 codes are not.

BCD Addition

BCD (8421) addition is performed by adding the digits in binary, starting at the least significant digit. If there is a carry out of one or the result is an illegal code, then add 6 (0110), and add the resulting carry to the next most significant digit.

Example 3-1: Add 647 to 492 in BCD (8421) code.

SOLUTION:

The problem in binary is:

0110   0100   0111
0100   1001   0010
_____

Adding the right group in binary:

$$
\begin{array}{r}
0111 \\
0010 \\
\hline
1001
\end{array}
$$

Since the carry out is zero and the code 1001 is permitted, we shall proceed to the middle group:

$$
\begin{array}{r}
0100 \\
1001 \\
\hline
1101
\end{array}
$$

There is no carry out but the result is an illegal code (1101 is not part of the 8421 code system). Therefore, add 6 (0110) to the result:

$$
\begin{array}{r}
1101 \\
0110 \\
\hline
1\ 0011
\end{array}
$$

We can now proceed to the MSB:

$$
\begin{array}{r}
1 \\
0110 \\
0100 \\
\hline
1011
\end{array}
$$

We must again add 6:

$$
\begin{array}{r}
1011 \\
0110 \\
\hline
1\ 0001
\end{array}
$$

The final solution is as follows:

647	0110	0100	0111
492	0100	1001	0010
1139	1 0001	0011	1001

Note that the binary work checks with the decimal result.

Example 3-2: Add 4318 and 7678 in 8421 BCD.

SOLUTION:

				1
4318	0100	0011	0001	1000
7678	0111	0110	0111	1000
	1011	1001	1001	0000
	0110	0000	0000	0110
11996	1 0001	1001	1001	0110

3-3 NONWEIGHTED CODES

Nonweighted codes are codes that are not positionally weighted. That is, each position within the binary number is not assigned a fixed value. We shall consider two such codes: excess-3 and Gray.

Excess-3 Code

Excess-3, also called XS3, is a nonweighted code used to express decimal numbers (Table 3-1). The code derives its name from the fact that each binary code is the corresponding 8421 code plus 0011 (3). Thus, 0 is 0000 + 0011, or 0011, and 9 is 1001 + 0011, or 1100. The code has some very interesting properties when used in addition. To add in XS3, add the binary numbers. If there is no carry out from the four-bit group, subtract 0011. If there is a carry out, add 0011.

Example 3-3: Add 3 and 2 in XS3.

SOLUTION:

$$
\begin{array}{ll}
3 = & 0110 \\
2 = & 0101 \\
\hline
\text{Sum} & 1011 \\
\text{Subtract} & 0011 \\
\hline
5 = & 1000 \\
\end{array}
$$

Example 3-4: Add 6 and 8 in XS3.

SOLUTION:

$$
\begin{array}{ll}
6 = & 1001 \\
8 = & 1011 \\
\hline
\text{Sum} & 1\ 0100 \\
\text{Add} & 0011 \\
\hline
14 = & 1\ 0111 \\
\end{array}
$$

Gray Codes

Consider a rotating disk that must provide an output of the position in three-bit binary (Fig. 3-1). When the brushes are on the black part, they output a 1. When on a white sector, they output a 0. However, consider what happens when the brushes are on the 111 sector and almost ready to enter the 000 sector. If one brush were slightly ahead of the other, say the 4's brush, the position would be indicated by a 011 instead of a 111 or 000. Therefore, a 180-degree (°) error in disk position would result. Since it is physically impossible to have all the brushes precisely aligned, some error would always be present at the edges of the sectors.

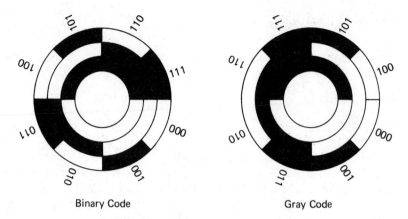

Binary Code   Gray Code

FIG. 3-1 Positional Indicating Codes

The Gray code was introduced to reduce this error. Shown in Table 3-2, it assures that only one bit will change each time the decimal number is incremented. Whereas the binary system requires all four bits to change when going from 7 to 8 (0111 to 1000), the Gray code requires only one bit to change (0100 to 1100).

A straight binary number can easily be converted to its Gray code equivalent:

A. Record the most significant bit.
B. Add this bit to the next position, recording the sum and neglecting any carry.
C. Record successive sums until completed.

TABLE 3-2. Gray Code

Decimal	Gray
0	0000
1	0001
2	0011
3	0010
4	0110
5	0111
6	0101
7	0100
8	1100
9	1101
10	1111
11	1110
12	1010
13	1011
14	1001
15	1000
16	1 1000

Example 3-5: Convert a binary 1011 to Gray code.

SOLUTION:

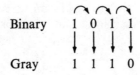

Binary   1   0   1   1

Gray   1   1   1   0

First record the 8's bit. Then add the 8's bit to the 4's bit $(1 + 0 = 1)$. Record it. Then add the 4's bit to the 2's bit $(0 + 1 = 1)$. Then add the 2's bit to the 1's bit $(1 + 1 = 10$; ignore carry). The result is 1110.

Example 3-6: Convert a binary 1001011 to Gray code.

SOLUTION:

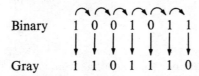

Binary   1   0   0   1   0   1   1

Gray   1   1   0   1   1   1   0

The Gray code can be converted to binary by a reverse process:

A. Record the most significant bit (MSB).
B. Add the binary MSB to the next significant bit of the Gray code.
C. Record the result, ignoring carries.
D. Continue the process until the LSB is reached.

Example 3-7: Convert the Gray code 1011 to binary.

SOLUTION:

Gray   1   0   1   1

Binary   1   1   0   1

First, record the MSB. Then add the MSB to the 4's bit of the Gray number $(1 + 0 = 1)$. Then add the 4's bit of the binary number to the 2's bit of the Gray code $(1 + 1 = 10$; ignore carry). Record the 0. Finally, add the 2's bit of the binary code to the 1's bit of the Gray code $(0 + 1 = 1)$ and record the sum.

Example 3-8: Convert a Gray 1001011 to binary.

SOLUTION:

Gray   1   0   0   1   0   1   1

Binary   1   1   1   0   0   1   0

This particular Gray code that we have used is only one of a family of such codes possessing this unique property. Others will be discussed further in Chapter 6.

Although you may feel that, having mastered a Gray code, you are ready for such color codes as orange, purple, and fuchsia, there are, unhappily, no such codes.

3-4 ERROR-DETECTING CODES

Because digital systems must be accurate to the digit, errors can become a serious problem. Several schemes have been devised to detect an error so that should one occur the binary word can either be ignored or retransmitted.

Parity

The simplest technique for detecting errors is that of adding an extra bit, known as a parity bit, to each word being transmitted. For odd parity this bit is set to a 1 or a 0 at the transmitter such that the sum of the 1 bits in the entire word is odd. The following are examples:

Parity	┌────	Data	────┐					Total 1's
1	1	0	1	0	1	1	0	5
0	1	0	1	0	1	1	1	5
1	0	0	1	0	0	1	0	3
1	0	0	0	0	0	0	0	1

Note that in each case the total number of 1 bits in the word (including the parity bit) is odd. One parity and seven data bits form the eight-bit word that is transmitted to the receiving location.

At the receiving location, each eight-bit word that is received is checked to see that the word contains an odd number of 1's. If a word is received that has an even number of 1's, the receiver will request a retransmission. Since most errors occur as a single-bit inversion, this system works quite well. However, should two errors occur within the word, they would remain undetected.

The above example was one of odd parity. Even parity is also used in many systems. This requires that the parity bit be set such that the sum of the 1's in the word is even. The following are examples:

Parity	┌────	Data	────┐					Total 1's
1	1	0	1	0	1	1	1	6
1	0	0	0	0	1	1	1	4
0	1	0	1	0	0	0	0	2

Check Sums

As discussed previously, simple parity will not detect two errors within the same word. For example, if 10101010 were transmitted using even parity and a 10011010 were received, it would appear as though no error had occurred. One way of overcoming this disadvantage is to use a sort of two-dimensional parity. As each word is transmitted, it is added to the previously sent word and sum retained at the transmitter. The following is an example:

Word A	1	0	1	1	0	1	1	1
Word B	0	0	1	0	0	0	1	0
Sum	1	1	0	1	1	0	0	1

Each successive word is added in this manner to the previous sum. At the end of the transmission, the sum (called a *check sum*) up to that time is sent. The receiver can check its sum against the transmitted sum. If the two are the same, then no errors were detected.

This is the type of transmission used in teleprocessing (TP) systems. Here, an entire message would be sent over a telephone transmission line, appearing as shown in Table 3-3. Summing the binary digits received, the result would be 0. If it is not, the entire message could again be sent.

TABLE 3-3. Check Sum

P	Data							Character
1	1	0	0	0	0	1	0	B
1	1	0	1	0	1	0	1	U
1	1	0	0	1	1	1	0	N
0	1	0	0	0	0	1	1	C
1	1	0	0	1	0	0	0	H
0	0	1	0	0	0	0	0	Space
0	1	0	0	1	1	1	1	O
0	1	0	0	0	1	1	0	F
0	0	1	0	0	0	0	0	Space
1	1	0	0	0	0	1	0	B
0	1	0	0	1	0	0	1	I
0	1	0	1	0	1	0	0	T
1	1	0	1	0	0	1	1	S
0	1	1	1	0	1	1	1	Check Sum

Parity Data Codes

By careful choice of codes, parity can be contained within each character. Two such codes are illustrated in Table 3-4. In each of these systems, only two 1's are contained in each character. The receiver can then check for this property in each

TABLE 3-4. Parity Data Codes

Decimal	2 Out of 5	Biquinary
0	00011	01 00001
1	00101	01 00010
2	00110	01 00100
3	01001	01 01000
4	01010	01 10000
5	01100	10 00001
6	10001	10 00010
7	10010	10 00100
8	10100	10 01000
9	11000	10 10000

character received. The 2 out of 5 code is an unweighted code that has been used in communications systems, whereas the biquinary code is used in the abacus.

3-5 ERROR-CORRECTING CODES

Because of the telephone company's great interest in communications, it has performed extensive research in finding codes that not only detect an error but correct that error. R. W. Hamming developed a system that is easily implemented. Assuming four data bits must be transmitted, the word format would be as follows:

$$D_7 \quad D_6 \quad D_5 \quad P_4 \quad D_3 \quad P_2 \quad P_1$$

where the D bits are the data bits and the P bits are the parity bits. P_1 is set so that it establishes even parity over bits 1, 3, 5, and 7 (P_1, D_3, D_5, and D_7). P_2 is set for even parity over bits 2, 3, 6, and 7 (P_2, D_3, D_6, and D_7). P_4 is set for even parity over bits 4, 5, 6, and 7 (P_4, D_5, D_6, and D_7).

Example 3-9: Data bits 1011 must be transmitted. Construct the even-parity, seven-bit Hamming code for this data.

SOLUTION:

P_1 must be a 1 in order for bits 1, 3, 5, and 7 to be even parity. P_2 must be a 0 in order for bits 2, 3, 6, and 7 to have even parity. P_4 must be a 0 in order for bits 4, 5, 6, and 7 to be even parity. Therefore, the final code is

$$
\begin{array}{ccccccc}
D_7 & D_6 & D_5 & P_4 & D_3 & P_2 & P_1 \\
1 & 0 & 1 & 0 & 1 & 0 & 1
\end{array}
$$

Example 3-10: Encode data bits 0101 into a seven-bit even-parity Hamming code.

SOLUTION:

$$
\begin{array}{ccccccc}
D_7 & D_6 & D_5 & P_4 & D_3 & P_2 & P_1 \\
0 & 1 & 0 & 1 & 1 & 0 & 1
\end{array}
$$

The Hamming-coded data are now ready for transmission and reception. At the receiver end, they are decoded to see if any errors occurred. Bits 1, 3, 5, and 7; bits 2, 3, 6, and 7; and bits 4, 5, 6, and 7 are all checked for even parity. Should they check out, there is no error. However, should there be an error, the problem bit can be located by forming a three-bit binary number out of the three parity checks. Assume the number 1011011 is received. Analyzing bits 4, 5, 6, and 7, of which P_4 is a parity bit, an error is detected. Thus, put a 1 in the 4's position of the error word. Bits 2, 3, 6, and 7, of which P_2 is the parity bit, contain no error, so put a 0 in the 2's position of the error word. Bits 1, 3, 5, and 7 associated with parity bit P_1 contain a parity error, so put a 1 in the 1's position of the error word. The resulting binary error word is 101, a decimal 5. Therefore, bit 5 of the transmitted code is in error. Since bit 5 was received as a 1, the corrected code should read 1001011.

Example 3-11: A seven-bit Hamming code is received as 1111101. What is the correct code?

SOLUTION:

$$D_7 \quad D_6 \quad D_5 \quad P_4 \quad D_3 \quad P_2 \quad P_1$$
$$1 \qquad 1 \qquad 1 \qquad 1 \qquad 1 \qquad 0 \qquad 1$$

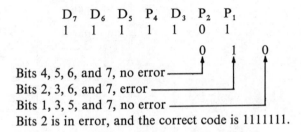

Bits 4, 5, 6, and 7, no error
Bits 2, 3, 6, and 7, error
Bits 1, 3, 5, and 7, no error
Bits 2 is in error, and the correct code is 1111111.

The illustrations above have shown a seven-bit Hamming code. However, the concept can be extended to any number of bits. A 15-bit code would have the following format:

$$D_{15} \quad D_{14} \quad D_{13} \quad D_{12} \quad D_{11} \quad D_{10} \quad D_9 \quad P_8 \quad D_7 \quad D_6 \quad D_5 \quad P_4 \quad D_3 \quad P_2 \quad P_1$$

Note that parity bits are inserted at each 2^n bit. This is true for Hamming codes of any length.

So far in this chapter, many different codes have been presented for representing numbers as binary codes. Next, alphanumeric representation using binary numbers will be examined.

3-6 ALPHANUMERIC CODES

Since computers, printers, and other devices must process both alphabetic and numeric information, several coding systems have been invented that represent this alphanumeric information as a series of 1's and 0's. These systems vary in complexity from the Morse code used in telegraph work to the Hollerith code used with punched cards.

Telegraph Codes

Codes representing letters and numbers have been used ever since the invention of the first practical telegraph system by Samuel F. B. Morse in 1844.

There are two code systems in practical use today: International Morse code and American Morse code. These are both illustrated in Table 3-5. Both are binary representations of alphanumeric characters. The international Morse code (also called *continental*) is used primarily for radio transmission. The American was used primarily for land line transmission using a "clacker" as a receiving device. The latter use has been replaced almost entirely with teletype transmission. Both systems were devised so the letters of the alphabet most commonly used have the simplest codes, requiring less time to transmit. Thus, the two most commonly used letters, E and T, have the simplest codes, "·" and "–," and the two least frequently used letters, Q and Z, have more complex codes, "– – · –" and "– – · ·," which take longer to transmit.

TABLE 3-5. Morse Code

	International	American		International	American
A	• –	• –	1	• – – – –	• – – •
B	– • • •	– • • •	2	• • – – –	• • – • •
C	– • – •	• • ·	3	• • • – –	• • • – •
D	– • •	– • •	4	• • • • –	• • • • –
E	•	•	5	• • • • •	– – –
F	• • – •	• – •	6	– • • • •	• • • • • •
G	– – •	– – •	7	– – • • •	– – • •
H	• • • •	• • • •	8	– – – • •	– • • • •
I	• •	• •	9	– – – – •	– • • –
J	• – – –	– • – •	0	– – – – –	————
K	– • –	– • –	.	• – • – • –	• • – – • •
L	• – • •	————	,	– – • • – –	• – • –
M	– –	– –	:	– – • • •	– • – • •
N	– •	– •	;	– • – • – •	• • • • •
O	– – –	• •	-	– • • • • –	• • • • • – • •
P	• – – •	• • • • •	!	– – • • –	– – – •
Q	– – • –	• • – •	'	• – – – – •	• • • – • • – • •
R	• – •	• • •	"	• – • • – •	• • – • • – •
S	• • •	• • •	/	– • • – •	• • – –
T	–	–	?	• • – – • •	– • • – •
U	• • –	• • –			
V	• • • –	• • • –			
W	• – –	• – –			
X	– • • –	• – • •			
Y	– • – –	• • • •			
Z	– – • •	• • • •			

Teletypewriter Code

Both civilian and military communications use a five-level (five-bit) teletypewriter code (Table 3-6). This code is an outgrowth of Morse codes, with the simpler binary numbers being assigned to the more common letters and more complex binary words being assigned to the infrequently used letters. There are two modes to the system: uppercase and lowercase. To send a "3," for example, the operator would first send a "figures" and then a "3." There are two configurations of machines: those that automatically return to the lowercase when a space is transmitted, and those that do not unshift upon transmitting a space.

The five-level code is not well suited for communications with computers, because each letter is not in an ascending binary sequence, making alphabetizing difficult. Both ASCII and EBCDIC overcome this difficulty.

ASCII Code

The American Standard Code for Information Interchange (ASCII, pronounced "as-kee") was developed out of a need for an orderly binary code as applied to alphanumeric data (Table 3-7). It is used extensively for printers and terminals that interface with small computer systems; many large systems also make provisions for its accommodation. Because characters are assigned in ascending binary numbers, ASCII is very easy for a computer to alphabetize and sort.

ASCII is basically a seven-bit code. An eighth bit is usually added and is (a) always set to a 1, (b) always set to a 0, or (c) used as a parity bit.

EBCDIC Code

The Extended Binary Coded Decimal Interchange Code (EBCDIC, pronounced "eb-si-dik" by the industry) (Table 3-8) is used by most large computers for communicating in alphanumeric data. Unlike ASCII, which uses a straight binary sequence for representing characters, this code uses binary-coded decimal as the basis of binary assignment. The entire system is closely related to punch card codes, making implementation of hardware quite simple.

Hollerith Code

The Hollerith code is the code used in punched cards. Prior to discussing the code itself, let us examine how a punched card is organized. Each card has 80 columns, oriented vertically, and 12 rows, oriented horizontally (Fig. 3-2). The rows are numbered 0 through 9, 11, and 12. The columns are numbered 1 through 80, each one containing one character. Each character is uniquely identified by the rows punched in that column. The letter A is a 12-1 punch (a punch in the 12 row and a punch in the 1 row), for example.

TABLE 3-6. Five-Level Teletypewriter Code

TABLE 3-7. American Standard Code for Information Interchange (ASCII)

Most Significant Digit (Hex)

		0	1	2	3	4	5	6	7
	0	NUL	DLE	SP	0	@	P	`	p
	1	SOH	DC1	!	1	A	Q	a	q
	2	STX	DC2	"	2	B	R	b	r
	3	ETX	DC3	#	3	C	S	c	s
	4	EOT	DC4	$	4	D	T	d	t
	5	ENQ	NAK	%	5	E	U	e	u
Least	6	ACK	SYN	&	6	F	V	f	v
Significant	7	BEL	ETB	'	7	G	W	g	w
Digit	8	BS	CAN	(	8	H	X	h	x
(LSD)	9	HT	EM	)	9	I	Y	i	y
	A	LF	SUB	*	:	J	Z	j	z
	B	VT	ESC	+	;	K	[	k	{
	C	FF	FS	,	<	L	\	l	¦
	D	CR	GS	−	=	M	]	m	}
	E	SO	RS	.	>	N	^	n	~
	F	SI	US	/	?	O	−	o	DEL

ACK	Acknowledge	FF	Form feed
BEL	Bell	FS	File Separator
BS	Backspace	GS	Group Separator
CAN	Cancel	HT	Horizontal tabulation
CR	Carriage return	LF	Line feed
DC1	Device control 1	NAK	Negative acknowledge
DC2	Device control 2	NUL	Null or all zeros
DC3	Device control 3	RS	Record Separator
DC4	Device control 4	SI	Shift in
DEL	Delete	SO	Shift out
DLE	Data link escape	SOH	Start of heading
EM	End of medium	SP	Space
ENQ	Enquiry	STX	Start of text
EOT	End of transmission	SUB	Substitute
ESC	Escape	SYN	Synchronous idle
ETB	End of transmission block	US	Unit Separator
ETX	End of text	VT	Vertical tabulation

The Hollerith code is BCD, rather than natural binary oriented. Thus, translation to and from EBCDIC is fairly simple. Since most large computers use punched cards, they use Hollerith for their card readers and punches and EBCDIC within the computer itself.

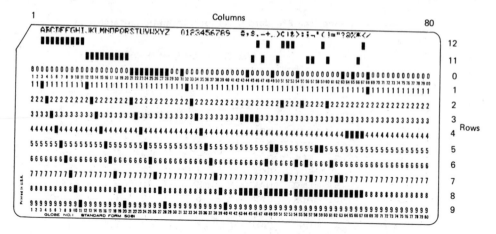

FIG. 3-2 Hollerith Code

TABLE 3-8. Extended Binary Coded Decimal Interchange Code (EBCDIC)

Most Significant Digit (Hex)

		0	1	2	3	4	5	6	7	8	9	A	B	C	D	E	F
	0	NUL	DLE	DS		SP	&							{	}	\	0
	1	SOH	DC1	SOS				/		a	j	~		A	J		1
	2	STX	DC2	FS	SYN					b	k	s		B	K	S	2
	3	ETX	DC3							c	l	t		C	L	T	3
	4	PF	RES	BYP	PN					d	m	u		D	M	U	4
	5	HT	NL	LF	RS					e	n	v		E	N	V	5
	6	LC	BS	EOB	UC					f	o	w		F	O	W	6
LSD	7	DEL	IL	PRE	EOT					g	p	x		G	P	X	7
(Hex)	8		CAN							h	q	y		H	Q	Y	8
	9		EM							i	r	z		I	R	Z	9
	A	SMM	CC	SM		¢	!	¦	:								
	B	VT				.	$	,	#								
	C	FF	IFS		DC4	<	*	%	@								
	D	CR	IGS	ENQ	NAK	(	)	–	'								
	E	SO	IRS	ACK		+	;	>	=								
	F	SI	IUS	BEL	SUB	|	¬	?	"								

3-7 SUMMARY

Both decimal numbers and alphabetic letters can be represented as binary codes. Decimal numbers can be expressed as binary-coded decimal numbers, simplifying their interpretation but requiring larger binary numbers. Where special additive requirements exist X53 can be used, and where mechanical systems must be used binary-coded Gray codes are helpful.

Several systems are being used today for error detection and correction. Biquinary codes use only two 1 bits in each word. Parity can be used with any code system for detection of single-bit errors and a check sum added for entire messages. Hamming codes not only detect errors but can also correct them.

Several systems have been used to represent alphanumeric data. At first, Morse codes were used. Later, the teletypewriter was introduced, and a five-level code was used. ASCII is used for scientific, binary-oriented machines and EBCDIC for business-oriented BCD machines. Hollerith is used for punched card data.

Thus far, you can now express a decimal 9 as follows:

Decimal	9	Natural BCD	1001
Binary	1001	XS3	1100
Octal	11	Gray	1101
Hexadecimal	9	Biquinary	10 10000
Five-level teletype	00011		
ASCII	0111001		
EBCDIC	11111001		
Hollerith	9		

In Chapter 4 we shall examine the mathematical basis of the hardware necessary for implementing these codes.

3-8 PROBLEMS

3-1. Express the following decimal numbers in 8421 BCD code:
   (a) 328   (b) 1497   (c) 9725

3-2. Express the following 8421 BCD numbers as decimals:
   (a) 1001 0101 0110 1000
   (b) 0110 1000 0001 0010
   (c) 1001 0111 0100 0101

3-3. Express the following decimal numbers in 2421 and 5211 codes:
   (a) 145   (b) 726   (c) 6725

3-4. Express the following 2421 numbers as a decimal:
   (a) 1111 1011 0100
   (b) 0010 1110 0001

3-5. Express the following 5211 numbers in decimal:
   (a) 1000 1100 0101 0001
   (b) 1110 1010 0011 1111

3-6. Express the following decimal numbers in XS3:
   (a) 124   (b) 7621   (c) 1643

3-7. Express the following XS3 numbers as decimals:
   (a) 0110 1011 1100 0111
   (b) 0011 0101 1010 0100

3-8. Add the following XS3 numbers, expressing the result in XS3.

(a) 0011 + 0100   (b) 0110 + 1010   (c) 0101 + 1001

3-9. Convert the following decimals to Gray code:

(a) 3   (b) 14   (c) 10

3-10. Convert the following binary numbers to Gray code:

(a) 10101   (b) 101101   (c) 1101011

3-11. Convert the following Gray codes to binary:

(a) 10i1   (b) 110101   (c) 101111

3-12. Using even parity, which of the following words contain an error?

(a) 10101010   (b) 11110110   (c) 10111001

(d) 10101111   (e) 11110111   (f) 10111101

3-13. What should the check sum be of the following characters?

$$
\begin{matrix}
0 & 1 & 0 & 1 & 1 & 0 & 1 & 0 \\
1 & 1 & 0 & 0 & 0 & 1 & 0 & 1 \\
1 & 1 & 0 & 1 & 1 & 0 & 0 & 1
\end{matrix}
$$

3-14. Convert the following decimal numbers to both 2 out of 5 and biquinary codes:

(a) 276   (b) 539   (c) 7658

3-15. Convert the following biquinary numbers to decimal:

(a) 0101000 1001000 0100100

(b) 0100001 1010000 0101000

(c) 1001000 1000010 0110000

3-16. Convert the following 2 out of 5 numbers to decimal:

(a) 10100 00110 11000

(b) 10010 01010 10100

(c) 00110 00011 11000

3-17. Encode the following binary words into seven-bit even-parity Hamming code:

(a) 1000   (b) 0101   (c) 1011

3-18. Detect and correct any errors in the following even-parity Hamming code words:

(a) 1010011   (b) 0101110   (c) 0010000

3-19. Encode the word "BINARY" into both international and American Morse code.

3-20. What is the message in Fig. 3-3?

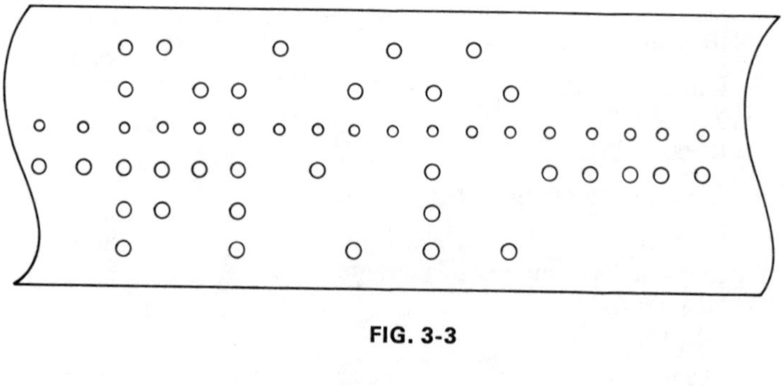

FIG. 3-3

3-21. What is the following ASCII-coded word? The numbers are in hex.

D4   C9   D0   AD   D4   CF   D0

3-22. What is the following EBCDIC-coded word? The numbers are in hexadecimal.

C6   C9   D9   E2   E3   40   C3   C1   D9   C4

3-23. Read the punched card in Fig. 3-4.

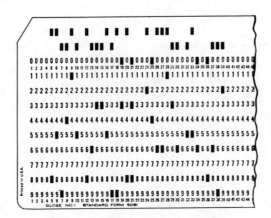

FIG. 3-4

BOOLEAN ALGEBRA

4

4-1 BINARY DECISIONS

We are faced every day with making logic decisions: "Should I emit an audible sound because I stubbed my toe or not?" "Should I count to 10 or not?" "Should I jump up and down or not?" Each of these decisions requires a yes or no answer; there are only two possible answers. Therefore, each of these decisions is a binary decision.

This same binary decision making also applies to formal logic. In the study of syllogisms, binary decisions are inferred. For example, consider the following:

> *All elephants waltz.*
> *Packy is an elephant.*
> *Therefore, Packy waltzes.*

The first statement, the major premise, states a positive fact. There are only two possibilities: Either all elephants waltz or all elephants do not waltz. Thus, this is a binary statement. Similarly, the second statement, the minor premise, also states a binary fact, for either Packy is an elephant or Packy is not an elephant. Again, the third statement is also a binary decision, for Packy either waltzes or Packy does not waltz. This system of reasoning is called logic, and the statements are called truth functions. The syllogism illustrates in rhetoric the operations that logic circuits perform in electronics.

4-2 DEVELOPMENT OF BOOLEAN ALGEBRA

Aristotle (384–322 B.C.) constructed a complete system of formal logic and wrote six famous works on the subject, contributing greatly to the organization of man's reasoning. For centuries afterward, mathematicians tried unsuccessfully to solve these logic problems using conventional algebra. It was not until 1854 that George Boole (1815–1864) developed a mathematical system of logic, expressing truth functions as symbols and then manipulating these symbols to arrive at a conclusion. His new system was not the algebra we all know and love from high school and college. His was a new system, an algebra of logic. For example, in his system, $A + A = A$, not $2A$ as in algebra.

Boole's work remained in the realm of philosophy until 1938 when Claude E. Shannon wrote a paper titled "A Symbolic Analysis of Relay Switching Circuits." In this paper he used Boole's algebraic system to solve relay logic problems. As rhetorical logic problems are binary decisions, true or false, relay problems are also binary, either energized or not energized. Lights are either on or off, pulses present or not present. Because of these two men, our system of Boolean algebra has been applied to the solution of any electronic circuit involving only two possible states.

We shall, in this chapter, deal with Boolean algebra, applying it to any logic circuit. The objectives of this use are as follows:

A. To simplify the procedure necessary in the solution of logical problems.
B. To simplify any circuit to its fewest components necessary to perform the function.

4-3 TRUTH FUNCTIONS

Consider the following statement: I want to buy a dog that has blue eyes and brown hair. The statement relates a desired output, the purchase of a dog, to two essential facts that must both be present: it must have both blue eyes and brown hair. We can analyze this statement further by forming a diagram, called a Venn diagram, to express all categories of dogs and their acceptability to me (Fig. 4-1). All dogs are represented in the diagram. Blue-eyed dogs are shown by circle A and brown-haired dogs by circle

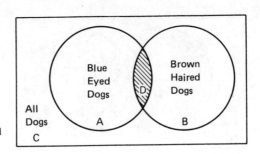

FIG. 4-1 Venn Diagram of All Dogs, AND Connective

TABLE 4-1. Truth Table of All Dogs, AND Connective

Brown haired	F	F	T	T
Blue eyed	F	T	F	T
Eligible	F	F	F	T

B. Area *C* represents those dogs who are neither blue eyed nor brown haired. The area of the intersection of the two circles, *D*, represents those eligible for purchase, for they are both blue eyed and brown haired. Note, however, that only those dogs within this shaded area would be acceptable. This type of relationship is called an **AND** connective. Both *A* **AND** *B* must be true for the desired output.

This relationship can also be expressed in another way, using a device called a truth table (Table 4-1). In this table, all four possible states are possible:

A. Those who have neither blue eyes nor brown hair have an F (false) after each attribute.

B. Those who have blue eyes (true) but not brown hair (false).

C. Those who are not blue eyed (false) but who are brown haired (true).

D. Those who are both blue eyed (true) and brown haired (true).

However, out of the four cases, in only one case is a dog eligible: when it is both blue eyed and brown haired. This is shown by the single T in the output term, "eligible." All other positions contain an F.

Now consider the following statement: I want to purchase a dog that is blue eyed or brown haired. Do you see the difference between this statement and the previous one? In this statement, either quality being present makes it eligible for purchase. Figure 4-2 illustrates a Venn diagram of this problem. Note that dogs possessing either quality are now eligible for purchase. Thus, all of those shown by the shaded part are eligible.

The truth table would appear as shown in Table 4-2. In this table, all possible states are examined and a decision made as to whether the output would be true or false. In three out of four cases a dog would be eligible for purchase. Only when it has neither blue eyes nor brown hair is it ineligible. Compare this **OR** connective with the **AND** connective shown in Table 4-1.

The truth tables perform a very important function in analysis of digital circuits. They enable the designer to examine each possible input combination and to define the output. A general model for this procedure is shown in Fig. 4-3. Given *n* possible

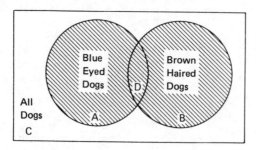

FIG. 4-2 Venn Diagram of All Dogs, **OR** Connective

TABLE 4-2. Truth Table of All Dogs, OR Connective

Brown haired	F	F	T	T
Blue eyed	F	T	F	T
Eligible	F	T	T	T

truth functions (variables), we must examine 2^n different combinations, defining the output state in each case. Assume, for example, that a burglar alarm is to be designed for a bank that will monitor three different things: whether the safe door is opened, whether the time is after 4 p.m., and whether the front door of the bank is opened. Each possible combination must be examined, as shown in Table 4-3.

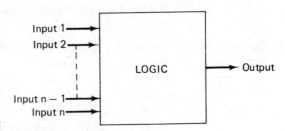

FIG. 4-3 General Model for a Digital Circuit

TABLE 4-3. Truth Functions for Bank Alarm

	0	1	2	3	4	5	6	7
Safe opened	F	F	F	F	T	T	T	T
After 4 p.m.	F	F	T	T	F	F	T	T
Front door opened	F	T	F	T	F	T	F	T
Alarm will sound	F	F	F	T	F	F	T	T

Note that the table shows every possible combination of the inputs. Since there are three possible input variables, there will be 2^3 or 8 possible combinations of those variables that must be examined. To assure that all are represented in the truth table, the F's and the T's are arranged in binary combinations representing the decimal numbers 0 through 7, with F's representing 0's and T's representing 1's. For example, consider the vertical column labeled "3". Reading from the top to the bottom of the column, an F, a T, and a T are encountered, forming the word FTT. When a 0 is substituted for the F and 1's for the T's, the binary word 011 emerges, which is the decimal 3. Thus, the column is labeled with a 3. All other columns are similarly labeled by the decimal equivalent of the binary numbers formed from the F's and the T's.

Many of the decisions involving truth functions are purely arbitrary. For example, consider when the safe is not opened, it is after 4 p.m., and the front door is opened. It might be that some of the employees are working late and have the front door opened. The designer must decide whether the alarm must ring under these conditions, using his own sense of reason.

Many of the conclusions we reach regarding things around us are the result of examining many cases and making generalizations about them. For example, inserting one's digit into a light socket for the first time can result in a shocking experience. However, inserting the digit for a second time could lead one to the conclusion that such activity always results in an unpleasant experience. Thus, we have generalized from specific cases, using a thought process called inductive reasoning. However, not all cases have been examined. It may be possible that if one were to insert one's digit into a light socket in the third floor of the Imperial Hotel, room 326, in Southern Chatteroy, Washington, on October 22, 1993, at 6:32 p.m. that nothing would happen. Thus, our induction was imperfect. However, in digital, since we draw conclusions only after examining all possible cases, we call our procedure *perfect induction*. Next, we shall examine the logic elements necessary for implementing this method.

4-4 INTRODUCTION TO BOOLEAN ALGEBRA

Boolean algebra is a system of mathematical logic. It differs from both ordinary algebra and the binary number system. As an illustration, in Boolean, $1 + 1 = 1$; in binary arithmetic the result is 10. Thus, although there are similarities, Boolean algebra is a unique system.

There are two constants within the Boolean system: 0 and 1. Every number is either a 0 or a 1. There are no negative or fractional numbers. Thus,

$$\text{If } X = 1, \quad \text{then } X \neq 0 \tag{4-1}$$

$$\text{If } X = 0, \quad \text{then } X \neq 1 \tag{4-2}$$

This is equivalent to saying "When you're hot, you're hot and when you're not, you're not."

4-5 THE AND OPERATOR

We use the **AND** operator every day. To put a car into motion, we must both start the engine **AND** engage the transmission. Notice the **AND** function. This is defined in Boolean algebra by use of the dot ($\cdot$), or no operator symbol at all. Thus, it is similar to multiplication in ordinary algebra. For example: $A \cdot B = C$ means that if A is true **AND** B is true, then C will be true. Under any other condition, C will be false. There are four different combinations of A and B that must be considered, as shown in Table 4-4. Note that only when both A and B are 1 (or true) will the output, C, be 1 (or true).

TABLE 4-4. AND Functions

$$A \cdot B = C$$

$$0 \cdot 0 = 0$$
$$0 \cdot 1 = 0$$
$$1 \cdot 0 = 0$$
$$1 \cdot 1 = 1$$

AND Gates

The **AND** function is also used in electronics. (Otherwise, why would you be reading this book?) Figure 4-4 shows the symbol for an **AND** gate and its associated truth table. This particular gate has two inputs and one output. However, prior to analysis, we must ask the question, "What voltage level is considered true and what level false?" Since many logic gates that are on the market today define a 0 V level as false and a +5 V level as true, we shall make this assumption in this example. Thus,

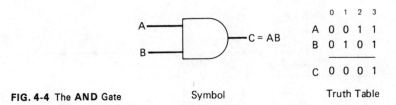

		0	1	2	3
	A	0	0	1	1
	B	0	1	0	1
	C	0	0	0	1

FIG. 4-4 The **AND** Gate   Symbol   Truth Table

this gate will provide +5 V on its output, *C*, if and only if both *A* and *B* inputs are at +5 V. Under any other set of input conditions, the output will be 0 V. Keep in mind that both the input and the output operate only at +5 V and ground. Ideally, no other voltages ever occur at either the inputs or the output of the gate. In actual practice, any voltage above a certain level (3 V, for example) is considered a logical true, and any voltage below a certain voltage (1 V, for example) is considered false. Thus, each input receives only a logical true (also called a logical 1) or a logical false (also called a logical 0).

AND gates may have any number of inputs. Figure 4-5 shows a four-input **AND** gate and its associated truth table. Note that the output will only be a 1 if all four inputs have 1's on them.

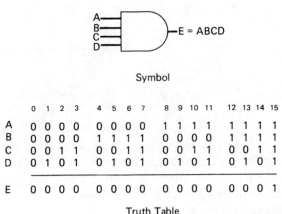

Symbol

	0	1	2	3	4	5	6	7	8	9	10	11	12	13	14	15
A	0	0	0	0	0	0	0	0	1	1	1	1	1	1	1	1
B	0	0	0	0	1	1	1	1	0	0	0	0	1	1	1	1
C	0	0	1	1	0	0	1	1	0	0	1	1	0	0	1	1
D	0	1	0	1	0	1	0	1	0	1	0	1	0	1	0	1
E	0	0	0	0	0	0	0	0	0	0	0	0	0	0	0	1

FIG. 4-5 The Four Input **AND** Gate   Truth Table

AND Circuits

It might be helpful at this point to present some possible physical realizations of the two-input **AND** gate. Figures 4-6, 4-7, and 4-8 show three such possible circuits that could be used as **AND** gates. Figure 4-6 shows a relay circuit. Plus five volts applied to both points A and B will cause relays K_1 and K_2 to energize, supplying $+5$ V to point C, the output, via the closed K_1 and K_2 contacts.

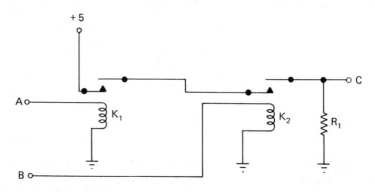

FIG. 4-6 Relay **AND** Gate

Figure 4-7 shows a semiconductor diode realization of the same **AND** gate. Point C will be $+5$ V only if both inputs A and B are at $+5$ V. If either A or B is at 0 V, the output will be 0 V (approximately).

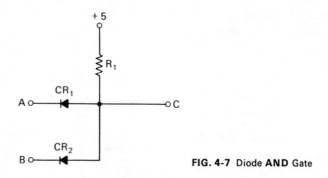

FIG. 4-7 Diode **AND** Gate

Figure 4-8 shows a transistor **AND** gate. When both A and B are at $+5$ V, transistors Q_1 and Q_2 conduct, moving point N to ground. This cuts off Q_3, driving point C to $+5$ V. However, if either A or B is at ground level, either Q_1 or Q_2 will be cut off, sending point N positive and providing base current to Q_3. Thus, point C will go to ground.

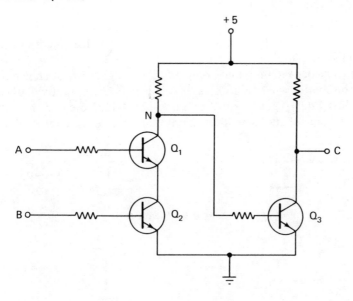

FIG. 4-8 Transistor **AND** Gate

AND Laws*

From the foregoing discussion there are three Boolean algebraic laws that will be examined closely:

$$A \cdot 1 = A \tag{4-3}$$

$$A \cdot 0 = 0 \tag{4-4}$$

$$A \cdot A = A \tag{4-5}$$

All three of these can be verified by remembering what the **AND** symbol means. Consider Eq. (4-3), and apply it to a two-input **AND** gate as shown in Fig. 4-9. If A equals 0 and the other input is 1, the output is 0 [Fig. 4-9(a)]. If A equals 1 and the

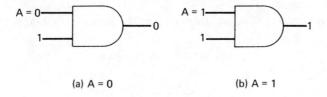

(a) A = 0   (b) A = 1

FIG. 4-9 Verifying $A \cdot 1 = A$

*Although not mathematically precise, all axioms, postulates, lemmas, and theorems are called laws in this chapter for simplicity and ease of discussion.

other input is 1, the output is 1 [Fig. 4-9(b)]. Thus, the output is always equal to the A input. Note that we could just as well replace the gate with a length of wire from the A input to the output.

Next, consider Eq. (4-4) and apply it to a two-input gate as shown in Fig. 4-10. Note that no matter what value A takes on, 1 or 0, the output will always be a 0.

Consider Eq. (4-5) and its realization in Fig. 4-11. Note that the output always takes on the value of A. These illustrations demonstrate the truth of the three laws.

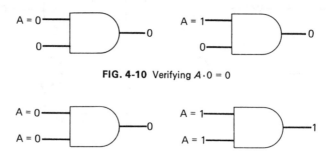

FIG. 4-10 Verifying $A \cdot 0 = 0$

FIG. 4-11 Verifying $A \cdot A = A$

4-6 THE **OR** OPERATOR

We use the **OR** operator all the time. "If either Mark is screaming or my wife is yelling, the house is in bedlam." "If I had a quarter or a half-dollar, I could buy a cup of coffee." "If I forget our anniversary or my wife's birthday, she gets angry." In each case, satisfying at least one of the conditions results in a definite effect. The **OR** operator is indicated by using a plus sign. Thus, $A + B = C$ means that if A is true **OR** B is true, then C will be true. Under any other set of conditions (there is only one other set of conditions—when both A and B are false), C will be false. There are four different combinations of A and B that must be examined in the light of the **OR** function, as shown in Table 4-5. Note that 1's appear at the output in three of the four cases. Compare this with the **AND** gate, which has a 1 in only one of the four different combinations.

This **OR** operator is called an *inclusive* **OR**, because it includes the case when

TABLE 4-5. **OR** Function

$A + B = C$
$0 + 0 = 0$
$0 + 1 = 1$
$1 + 0 = 1$
$1 + 1 = 1$

both inputs are true. Whenever an **OR** function is mentioned, the inclusive **OR** is what is meant. If I were to forget both our anniversary and my wife's birthday, she would, indeed, be angry. This is definitely an inclusive **OR**.

The other type of **OR** function is called an exclusive **OR**. "I will drive either the truck or the car to work." This is an exclusive **OR**, because it is impossible for me to drive both simultaneously.

OR Gates

The symbol for the **OR** gate as used in electronics is shown in Fig. 4-12, along with its associated truth table. Assuming $+5$ V is a logical 1 and 0 V is a logical 0, the **OR** gate can be described as a device that outputs a logical 1 when a logical 1 is present on any of the inputs. The **OR** gate may have any number of inputs. A four-input gate is shown in Fig. 4-13, along with its truth table. Note that any input being a 1 causes the output to be a 1.

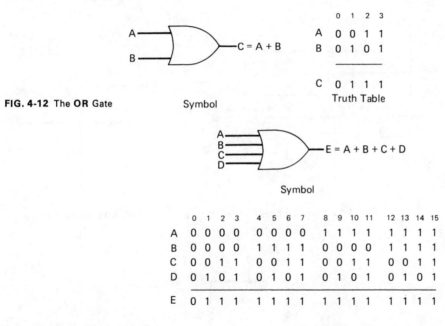

FIG. 4-12 The **OR** Gate   Symbol

	0	1	2	3
A	0	0	1	1
B	0	1	0	1
C	0	1	1	1

Truth Table

$C = A + B$

$E = A + B + C + D$

Symbol

	0	1	2	3	4	5	6	7	8	9	10	11	12	13	14	15
A	0	0	0	0	0	0	0	0	1	1	1	1	1	1	1	1
B	0	0	0	0	1	1	1	1	0	0	0	0	1	1	1	1
C	0	0	1	1	0	0	1	1	0	0	1	1	0	0	1	1
D	0	1	0	1	0	1	0	1	0	1	0	1	0	1	0	1
E	0	1	1	1	1	1	1	1	1	1	1	1	1	1	1	1

FIG. 4-13 The Four-Input **OR** Gate   Truth Table

Exclusive OR Gates

Figure 4-14 shows the symbol and truth table for an exclusive **OR** gate. If either input, but not both, is true, the output is true. Under any other condition, the output is false.

A ⟩⟩⟩──── C = A ⊕ B

B

	0	1	2	3
A	0	0	1	1
B	0	1	0	1
C	0	1	1	0

Symbol   Truth Table   FIG. 4-14 The Exclusive OR Gate

OR Circuits

Figures 4-15, 4-16, and 4-17 show several possible circuits that will behave in the manner described by the **OR** gate. Figure 4-15 shows a relay **OR** gate realization. If either A or B is $+5$ V, one of the relay contacts (which are wired in parallel) will close, applying $+5$ V to point C.

Figure 4-16 illustrates a diode **OR** gate. Applying $+5$ V to input A, for example, will forward-bias CR_1, causing point C to go to $+5$ V.

Figure 4-17 illustrates a possible transistor **OR** gate. Applying $+5$ V to point

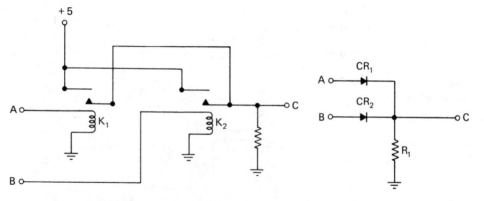

FIG. 4-15 Relay OR Gate   FIG. 4-16 Diode OR Gate

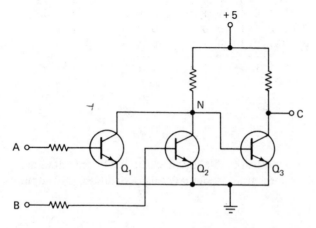

FIG. 4-17 Transistor OR Gate

A will cause Q_1 to conduct, causing point *N* to go to ground. This, in turn, will cut off Q_3, causing point *C* to go to +5 V. Applying +5 V to point *B* will cause Q_2 to conduct, resulting in the output again going to +5 V. If both inputs are grounded, Q_1 and Q_2 will cut off, causing point *N* to go positive, supplying current to the base of Q_3. This results in output *C* going to ground. Therefore, this circuit satisfies the definition given above for the **OR** gate.

OR Laws

As with the **AND** gate, there are several **OR** laws that become apparent by studying the **OR** gate:

$$A + 1 = 1 \tag{4-6}$$

$$A + 0 = A \tag{4-7}$$

$$A + A = A \tag{4-8}$$

The two possible cases of Eq. (4-6) are shown in Fig. 4-18. Regardless of what value *A* takes on, the output is always 1.

Figure 4-19 illustrates Eq. (4-7). When *A* is set to 0, the output is 0. When *A* is set to 1, the output is 1. Therefore, the output assumes the value of *A*.

Figure 4-20 illustrates Eq. (4-8). With *A* set to 0, the output is 0. With *A* set to 1, the output is 1. Thus, the output always equals *A*.

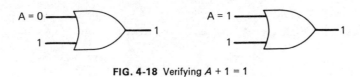

FIG. 4-18 Verifying *A* + 1 = 1

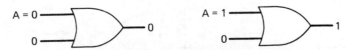

FIG. 4-19 Verifying *A* + 0 = *A*

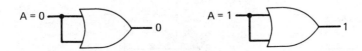

FIG. 4-20 Verifying *A* + *A* = *A*

4-7 THE **NOT** OPERATOR

The **NOT** operator changes the sense of an argument. Consider the following statements:

If we are at war, we do not have peace.
If we are not at war, we have peace.

The "not" used in the above sentences is an example of the **NOT** operator. Instead of having peace, we do not have peace. Instead of having war, we do not have war.

The symbol for the **NOT** operator is the vinculum, also called an overscore or bar. A symbolic representation of the above two sentences would be

$$W = \bar{P} \quad \text{and} \quad \bar{W} = P$$

Inverters

The logic diagrams for the **NOT** operator are called inverters and are shown with the truth table in Fig. 4-21. The circle actually represents the inversion and the triangle an amplifier. When a single circuit is used for inversion alone the triangle is included with the symbol. When the circuit is used for both gating and inversion the circle is shown in series with a gate input or output lead. Some examples of this use of the circle are shown in Fig. 4-22, along with their meanings. In each case the circle represents an inversion.

	0	1
A	0	1
B	1	0

Truth Table

Symbols

FIG. 4-21 The Inverter

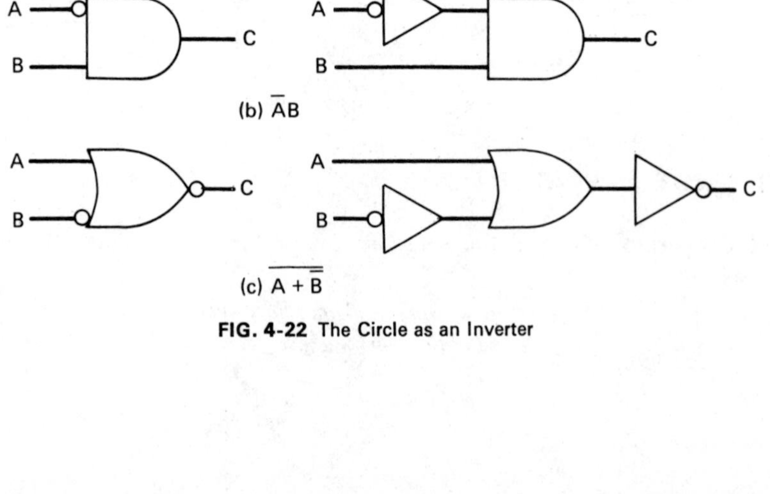

SYMBOL   MEANING

(a) $\overline{A + B}$

(b) $\overline{A}B$

(c) $\overline{\overline{A} + B}$

FIG. 4-22 The Circle as an Inverter

Inverting Circuits

Inverting circuits of two breeds are shown in Figs. 4-23 and 4-24. With $+5$ V applied to input A of the relay circuit, the relay will energize, opening the normally closed contacts, causing 0 V to appear at the output lead, B. In the transistor circuit (Fig. 4-24), $+5$ V applied to input A will turn the transistor on, causing it to conduct, resulting in a ground level at point B. With a ground level applied to point A, the transistor will turn off, resulting in $+5$ V at output B.

Note that an inverter cannot be designed using passive devices alone, such as resistors and diodes.

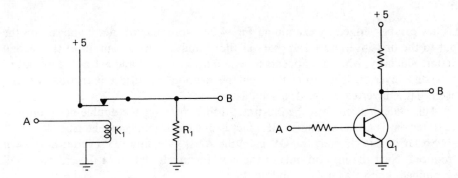

FIG. 4-23 Relay Inverter   **FIG. 4-24** Transistor Inverter

NOT Laws

There are several laws of Boolean algebra that become apparent when examining the inverter:

$$\bar{0} = 1 \tag{4-9}$$

$$\bar{1} = 0 \tag{4-10}$$

$$\text{If } A = 0, \quad \text{then } \bar{A} = 1 \tag{4-11}$$

$$\text{If } A = 1, \quad \text{then } \bar{A} = 0 \tag{4-12}$$

Equations (4-9) through (4-12) can be verified by examining Fig. 4-25. There are only two possibilities for input A on the inverter: either the input is a 0 or it is a 1. If the input is 0, the output, A, must be 1. If the input is 1, the output, A, must be 0.

Since Boolean algebra originally came from rhetorical logic, these equations can also be verified by reason. Equations (4-9) through (4-12) would appear as follows:

FIG. 4-25 Verifying **NOT** Relationships

If a statement is not false, it must be true.

If a statement is not true, it must be false.

If a statement is false, then the negation of that statement is true.

If a statement is true, then the negation of that statement must be false.

A fourth law that emerges from the definition of an inverter is

$$\bar{\bar{A}} = A \tag{4-13}$$

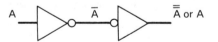

FIG. 4-26 Verifying $\bar{\bar{A}} = A$

This law can be verified by examining Fig. 4-26. Note that when a 0 appears on the input to the first inverter, a 1 appears at the $\bar{A}$ and a 0 at the output of the second inverter. Similarly, when a 1 appears at A, a 0 appears at $\bar{A}$ and a 1 at the output of the second inverter. Therefore, through the method of perfect induction, we can reason that A inverted twice, $\bar{\bar{A}}$, is identical to A.

This statement can also be illustrated using rhetoric: the double negation of a true statement is a true statement. If a fact is not not true, it must be true.

So far in this chapter, the **OR** gate, the **AND** gate, and the inverter have been introduced. Next the ins and outs of the laws for manipulation of these devices will be examined, known as in-laws and out-laws.

4-8 LAWS OF BOOLEAN ALGEBRA

Boolean algebra is a system of mathematics. As with any such system, there are fundamental laws that are used to build a workable, cohesive framework upon which are placed the theorems proceeding from these laws. Many of these laws have already been discussed in previous sections of this chapter. They will be repeated here, along with others that will provide all the tools necessary for manipulating Boolean expressions.

Laws of Complementation

The term *complement* simply means to invert, to change 1's to 0's and 0's to 1's. The five laws of complementation were discussed in Sec. 4-7, where inverters were introduced. They are as follows:

LAW 1	$\bar{0} = 1$
LAW 2	$\bar{1} = 0$
LAW 3	If $A = 0$, then $\bar{A} = 1$

LAW 4   If $A = 1$,   then $\bar{A} = 0$

LAW 5 $\bar{\bar{A}} = A$

AND Laws

The four **AND** laws are

LAW 6 $A \cdot 0 = 0$

LAW 7 $A \cdot 1 = A$

LAW 8 $A \cdot A = A$

LAW 9 $A \cdot \bar{A} = 0$

Laws 6 through 8 were discussed in Sec. 4-5 in the discussion of **AND** gates. Figure 4-27 illustrates Law 9. If A were 0, then $\bar{A}$ would be 1. The **AND** gate would then have a 0 on one of its inputs and a 1 on the other input, causing a 0 to appear at the output. If A were a 1, $\bar{A}$ would be a 0, and the **AND** gate would still output a 0. Thus, regardless of the value of A, $A \cdot \bar{A} = 0$.

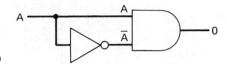

FIG. 4-27 Verifying $A \cdot \bar{A} = 0$

OR Laws

The four **OR** laws are as follows:

LAW 10 $A + 0 = A$

LAW 11 $A + 1 = 1$

LAW 12 $A + A = A$

LAW 13 $A + \bar{A} = 1$

Laws 10 through 12 were discussed in Sec. 4-6 when **OR** gates were considered. Law 13 is illustrated in Fig. 4-28. If A were a 1, then $\bar{A}$ would be 0. Therefore the expression of the **OR** gate would be $1 + 0 = 1$. If A were a 0, then $\bar{A}$ would be a 1. The **OR** gate function would receive $0 + 1$, and the output would be a 1. As can be seen, regardless of the value of A, $A + \bar{A} = 1$.

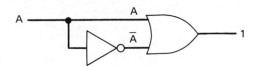

FIG. 4-28 Verifying $A + \bar{A} = 1$

Commutative Laws

The commutative laws allow the change in position of an **AND** or **OR** variable:

LAW 14 $\qquad\qquad\qquad A + B = B + A$

LAW 15 $\qquad\qquad\qquad A \cdot B = B \cdot A$

Table 4-6 is a truth table illustrating Law 14. By perfect induction, every case of $A + B$ yields an identical result to $B + A$. Thus, this law has been verified.

Table 4-7 illustrates a truth table for A, B, $A \cdot B$, and $B \cdot A$. Note that, after all cases have been examined, $A \cdot B = B \cdot A$.

TABLE 4-6. Verifying A + B = B + A

	0	1	2	3
A	0	0	1	1
B	0	1	0	1
$A + B$	0	1	1	1
$B + A$	0	1	1	1

TABLE 4-7. Verifying A·B = B·A

	0	1	2	3
A	0	0	1	1
B	0	1	0	1
$A \cdot B$	0	0	0	1
$B \cdot A$	0	0	0	1

Associative Laws

The associative laws allow the grouping of variables. The two associative laws are

LAW 16 $\qquad\qquad A + (B + C) = (A + B) + C$

LAW 17 $\qquad\qquad A \cdot (B \cdot C) = (A \cdot B) \cdot C$

Table 4-8 is a truth table for Law 16. **OR**ing A and B yields the result shown by $(A + B)$. **OR**ing B and C yields the result shown by $(B + C)$. The next step is to **OR** the row $(A + B)$ with row C, yielding the results shown by row $(A + B) + C$. Note that this result is in every case the same as **OR**ing row A with row $(B + C)$. This verifies the law.

TABLE 4-8. Verifying A + (B + C) = (A + B) + C

	0	1	2	3	4	5	6	7
A	0	0	0	0	1	1	1	1
B	0	0	1	1	0	0	1	1
C	0	1	0	1	0	1	0	1
$(A + B)$	0	0	1	1	1	1	1	1
$(B + C)$	0	1	1	1	0	1	1	1
$(A + B) + C$	0	1	1	1	1	1	1	1
$A + (B + C)$	0	1	1	1	1	1	1	1

Table 4-9 is a truth table illustrating Law 17. Again, the results of $(A \cdot B) \cdot C$ are, in every case, identical with the results of $A \cdot (B \cdot C)$. This verifies the law.

TABLE 4-9. Verifying $(A \cdot B) \cdot C = A \cdot (B \cdot C)$

	0	1	2	3	4	5	6	7
A	0	0	0	0	1	1	1	1
B	0	0	1	1	0	0	1	1
C	0	1	0	1	0	1	0	1
$(A \cdot B)$	0	0	0	0	0	0	1	1
$(B \cdot C)$	0	0	0	1	0	0	0	1
$(A \cdot B) \cdot C$	0	0	0	0	0	0	0	1
$A \cdot (B \cdot C)$	0	0	0	0	0	0	0	1

Distributive Laws

The distributive laws allow the factoring or multiplying out of expressions. Three distributive laws will be considered:

LAW 18 $A \cdot (B + C) = (A \cdot B) + (A \cdot C)$

LAW 19 $A + (B \cdot C) = (A + B) \cdot (A + C)$

LAW 20 $A + (\bar{A} \cdot B) = A + B$

Table 4-10 is a perfect induction proof for Law 18, since, in every case, the left-hand side of the identity is the same as the right-hand side.

TABLE 4-10. Verifying $A \cdot (B + C) = (A \cdot B) + (A \cdot C)$

	0	1	2	3	4	5	6	7
A	0	0	0	0	1	1	1	1
B	0	0	1	1	0	0	1	1
C	0	1	0	1	0	1	0	1
$(B + C)$	0	1	1	1	0	1	1	1
$(A \cdot B)$	0	0	0	0	0	0	1	1
$(A \cdot C)$	0	0	0	0	0	1	0	1
$A \cdot (B + C)$	0	0	0	0	0	1	1	1
$(A \cdot B) + (A \cdot C)$	0	0	0	0	0	1	1	1

Law 19 differs from the algebra we struggled with in high school. This Boolean law can be proved as follows:

$$A + BC = A + BC$$

$$= A \cdot 1 + BC \qquad \text{Law 7}$$

$$= A(1 + B) + BC \qquad \text{Laws 11 and 14}$$

$$= A + AB + BC \qquad \text{Law 18}$$

$$= A(1 + C) + AB + BC \qquad \text{Law 11}$$

$$= AA + AC + AB + BC \qquad \text{Laws 8 and 18}$$

$$= A(A + C) + BA + BC \qquad \text{Laws 18 and 15}$$

$$= A(A + C) + B(A + C) \qquad \text{Law 18}$$
$$= (A + C)A + (A + C)B \qquad \text{Law 15}$$
$$= (A + C)(A + B) \qquad \text{Law 18}$$
$$A + BC = (A + B)(A + C) \qquad \text{Law 15}$$

Law 20 is also a departure from numerical algebra. It can be proved as follows:

$$A + \bar{A}B = A + \bar{A}B$$
$$= A \cdot 1 + \bar{A}B \qquad \text{Law 7}$$
$$= A(1 + B) + \bar{A}B \qquad \text{Laws 11 and 14}$$
$$= A \cdot 1 + AB + \bar{A}B \qquad \text{Law 18}$$
$$= A + AB + \bar{A}B \qquad \text{Law 7}$$
$$= A + BA + B\bar{A} \qquad \text{Law 15}$$
$$= A + B(A + \bar{A}) \qquad \text{Law 18}$$
$$= A + B \cdot 1 \qquad \text{Law 13}$$
$$A + \bar{A}B = A + B \qquad \text{Law 7}$$

De Morgan's Theorem

One of the most powerful identities used in Boolean algebra is De Morgan's theorem. It provides two tools to the designer:

A. It allows removal of individual variables from under a **NOT** sign. For example, $\overline{A + BC}$ can be transformed into $\bar{A}(\bar{B} + \bar{C})$.

B. It allows transformation from a sum-of-products form to a product-of-sums form. For example, $A\bar{B}C + A\bar{B}\bar{C}$ can be transformed into

$$\overline{(\bar{A} + B + \bar{C})(\bar{A} + B + C)}$$

The theorem can be expressed by the following identity:

Law 21 $\qquad\qquad\qquad \bar{A} \cdot \bar{B} = \overline{A + B}$

To verify the identity, refer to Table 4-11. Note that for each assigned value of A and B the theorem is true. Therefore, the identity has been proved by perfect induction.

All of this is very interesting, but, as the monkey said to the dog concerning the dog's tail, "What good is it?" Let us examine the hardware implementation for each side of De Morgan's theorem. Figure 4-29 compares the logic required for each. Observe that the basic logic function involved can be either an **OR** gate or an **AND** gate, depending on what is available to the designer. Both forms execute the same function. This allows him to use two tools for implementing any function rather than one, and he can convert it to whatever form is convenient to him.

TABLE 4-11. Truth Table, De Morgan's Theorem

	0	1	2	3
A	0	0	1	1
B	0	1	0	1
$\bar{A}$	1	1	0	0
$\bar{B}$	1	0	1	0
$\bar{A} \cdot \bar{B}$	1	0	0	0
$A + B$	0	1	1	1
$\overline{A + B}$	1	0	0	0

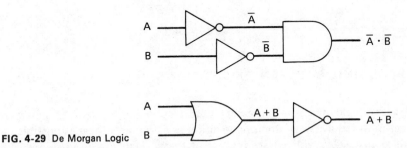

FIG. 4-29 De Morgan Logic

Although the identity above represents De Morgan's theorem, the transformation is more easily performed by following these steps:

A. Complement the entire function.

B. Change all the **ANDs** to **ORs** and all the **ORs** to **ANDs**.

C. Complement each of the individual variables.

This procedure is called *demorganization*.

Example 4-1: Demorganize the function $\overline{A\bar{B} + C}$.

SOLUTION:

Given	$\overline{A\bar{B} + C}$
Complement function	$A\bar{B} + C$
Change operators	$(A + \bar{B})(C)$
Complement variables	$(\bar{A} + B)(\bar{C})$

A second method of performing demorganization obeys the itty bitty ditty:

"Break the line, change the sign."

For example, assume we wish to demorganize $A + \overline{BC}$. We can break the line (the overscore) between B and C and change the operation sign from that of ANDing to that of ORing. This gives:

$$A + \bar{B} + \bar{C}$$

This method is especially useful in multiple levels of transformations.

Example 4-2: Reduce the expression: $A + B(\overline{C + \overline{DE}})$

SOLUTION:

We can first break the lower line between D and E yielding:

$$A + B(\overline{C + \bar{D} + \bar{E}})$$

Next, we can break the lines between C and D, and between D and E, changing the operations signs of each. This yields:

$$A + B(\bar{C}DE)$$

$$= A + B\bar{C}DE$$

4-9 REDUCING BOOLEAN EXPRESSIONS

Since every logic operator represents a corresponding element of hardware, the designer must reduce every Boolean equation to as simple a form as possible in order to reduce cost. This will require the use of the 21 laws discussed in Sec. 4-8. The techniques used for these reductions are similar to those used in ordinary algebra. The following procedure can be used as a general approach:

A. Multiply all variables necessary to remove parentheses.
B. Look for identical terms. Using Law 12, one of these can be dropped.
C. Look for a variable and its negation in the same term. This term can be dropped. For example,

$$A\bar{A}C = 0 \cdot C \qquad \text{Law 9}$$

$$= 0 \qquad \text{Law 6}$$

D. Look for pairs of terms that are identical except for one variable. If the one variable is missing, the larger term can be dropped. For example,

$$ABCD + ABD = ABD(C + 1) \qquad \text{Law 18}$$

$$= ABD \cdot 1 \qquad \text{Law 11}$$

$$= ABD \qquad \text{Law 7}$$

If the one variable is present but negated in the second term, it can be reduced. For example,

$$ABCD + A\bar{B}CD = ACD(B + \bar{B}) \qquad \text{Law 18}$$

$$= ACD \cdot 1 \qquad \text{Law 13}$$

$$= ACD \qquad \text{Law 7}$$

The preceding procedure can be used for most cases. However, each expression must be examined for combinations that would permit reduction.

Example 4-3: Reduce the expression $\overline{\overline{AB} + \bar{A}} + AB$.

SOLUTION:

Demorganize $\overline{AB}$	$\overline{\bar{A} + \bar{B} + \bar{A}} + AB$	Law 21
Reduce	$\overline{\bar{A} + \bar{B}} + AB$	Law 12
Reduce	$\overline{\bar{A} + \bar{B} + A}$	Law 20
Rearrange	$\overline{A + \bar{A} + \bar{B}}$	Law 14
Reduce	$\overline{1 + \bar{B}}$	Law 13
Reduce	$\bar{1}$	Law 11
Convert	0	Law 2

Example 4-4: Reduce the expression $AB + \overline{AC} + A\bar{B}C(AB + C)$.

SOLUTION:

Multiply	$AB + \overline{AC} + AAB\bar{B}C + A\bar{B}CC$	Law 18
Reduce	$AB + \overline{AC} + A\bar{B}C$	Laws 6, 8, 9
Demorganize $\overline{AC}$	$AB + \bar{A} + \bar{C} + A\bar{B}C$	Law 21
Rearrange	$AB + \bar{C} + \bar{A} + A\bar{B}C$	Law 14
Reduce	$AB + \bar{C} + \bar{A} + \bar{B}C$	Law 20
Rearrange	$\bar{A} + AB + \bar{C} + \bar{B}C$	Law 14
Reduce	$\bar{A} + B + \bar{C} + \bar{B}$	Law 20
Reduce	1	Laws 11, 13

Example 4-5: Reduce $\overline{\overline{A\bar{B} + ABC} + A(B + A\bar{B})}$.

SOLUTION:

Factor	$\overline{\overline{A(\bar{B} + BC)} + A(B + A\bar{B})}$	Law 18
Reduce	$\overline{\overline{A(\bar{B} + C)} + A(B + A)}$	Law 20
Multiply	$\overline{\overline{A\bar{B} + AC} + AA + AB}$	Law 18
Reduce	$\overline{\overline{A\bar{B} + AC} + A}$	Laws 7, 8, 11, 18
Demorganize $\overline{A\bar{B} + AC}$	$\overline{(\bar{A} + B)(\bar{A} + \bar{C}) + A}$	Law 21
Multiply	$\overline{\bar{A}\bar{A} + \bar{A}\bar{C} + \bar{A}B + B\bar{C} + A}$	Law 18
Reduce	$\overline{\bar{A} + \bar{A}\bar{C} + \bar{A}B + B\bar{C} + A}$	Law 8
Factor	$\overline{\bar{A}(1 + \bar{C} + B) + B\bar{C} + A}$	Law 18
Reduce	$\overline{\bar{A} + B\bar{C} + A}$	Laws 7, 11
Reduce	$\bar{1}$	Law 13
Reduce	0	Law 2

4-10 BOOLEAN EXPRESSIONS AND LOGIC DIAGRAMS

Boolean algebra is useless unless it can be translated into hardware, in the form of **AND** gates, **OR** gates, and inverters. Similarly, Boolean algebra can be a useful technique for analyzing existing circuits only if the hardware can be translated into a Boolean expression. Although this capability develops with experience, the next few paragraphs point in the right direction.

Algebra to Logic

The easiest way to convert an expression into a logic diagram is to start with the output and work toward the input. Assume the expression $\overline{A + \bar{B}C}$ is to be implemented in logic. Referring to Fig. 4-30, start with the final expression. Since this is basically a **NOT**ed function, it must be the inverse of $A + \bar{B}C$. Thus, show an inverter with an input expression $A + \bar{B}C$. Note that the circle is on the output, indicating that when $A + \bar{B}C$ is high, $\overline{A + \bar{B}C}$ is low. The circle should be drawn to indicate the location of the **NOT**ed function.

Next, consider the function on the input of the inverter, $A + \bar{B}C$. This is an **OR** of the terms A and $\bar{B}C$, so draw an **OR** gate with two inputs. One input is term A, and the other, term $\bar{B}C$.

The $\bar{B}C$ term is an **AND** function of $\bar{B}$ and C. Therefore, draw an **AND** gate with two inputs. Label the inputs B and C.

The $\bar{B}$ term can be left as shown, or another inverter can be shown with a B input.

Many inputs to logic systems are similar to the $\bar{B}$ input and enter the system

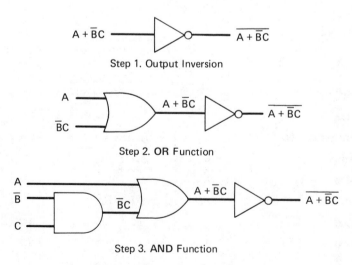

Step 1. Output Inversion

Step 2. OR Function

Step 3. AND Function

FIG. 4-30 Converting Boolean to Logic

as a 0 V signal when an event occurs. In this case, a *B* signal never exists in hardware. Consider Fig. 4-31, and assume the switch is ON when it is closed and OFF when open. In this case, the signal is true when the switch outputs a ground signal. If a +5 V signal were required to indicate the switch is ON, the $\bar{B}$ signal would have to feed an inverter whose output would be *B*. Thus, the **NOT** over the *B* indicates two things: (a) that the signal is true when a ground appears on the wire, and (b) that the signal must be inverted to obtain a high-level true signal.

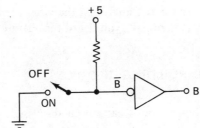

FIG. 4-31 Switching to Ground

Logic to Algebra

The reverse process can be used for converting logic to algebra. Instead of proceeding from output to input, start with the input signals and develop terms until the output is reached.

Consider the logic diagram shown in Fig. 4-32. First, label all the inputs. Observe

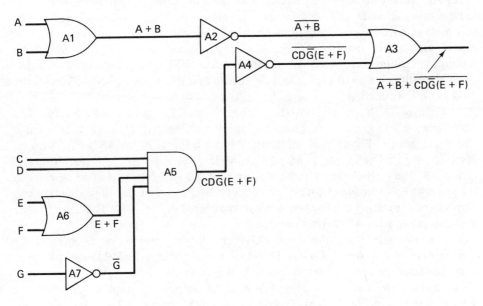

FIG. 4-32 Converting Logic to Boolean Algebra

that A and B feed **OR** gate $A1$. The output is therefore an **OR** of A and B, expressed as $A + B$. Similarly, E and F feed $A6$, resulting in $E + F$ at its output. G inverted is $\bar{G}$. **AND** gate $A5$ has inputs of C, D, $E + F$, and $\bar{G}$. Therefore, **AND**ing these terms yields $CD\bar{G}(E + F)$ at $A5$ output. This signal is inverted, yielding $\overline{CD\bar{G}(E + F)}$. Note that the entire expression is inverted by $A6$, and not the individual variables.

Similarly, $A2$ inverts $A + B$ to $\overline{A + B}$, providing inputs to $A3$ of $\overline{A + B}$ and $\overline{CD\bar{G}(E + F)}$. **OR**ing these with **OR** gate $A3$ gives the final result, $\overline{A + B} + \overline{CD\bar{G}(E + F)}$.

Using this procedure and analyzing the circuit from input to output, the final expression can easily (where have you heard that before?) be obtained.

4-11 UNIVERSAL BUILDING BLOCKS

Thus far, three building blocks have been discussed: **AND** gates, **OR** gates, and inverters. Wouldn't it be great if one unit could perform all three functions? There are, in fact, two such devices that can be used as universal building blocks: the **NAND** gate and the **NOR** gate. In this section we shall discuss these two devices and illustrate how the designer can convert from **AND/OR**/invert logic to **NAND** or **NOR** logic.

NAND Logic

A **NAND** gate is a piece of hardware. It is a particular, purchasable piece of hardware, usually an integrated circuit. This piece of hardware can perform all three logic functions: **AND, OR,** and invert. The reason this point is being made is that this piece of hardware is many times confused with the function it is performing. A **NAND** gate can be used to perform an **OR** function. Thus, out of one side of my mouth I am calling it a **NAND** gate (**NAND** means **NOT AND**), and out of the other side I am calling it an **OR** gate. The difference is that it is a piece of **NAND** hardware used as an **OR** function.

Figure 4-33 illustrates a **NAND** gate circuit, with its associated truth table. The 1's indicate $+5$ V, and the 0's, ground. Note that the output, C, is low only when A and B are both 1. Figure 4-34 illustrates this logic function, an **AND** gate with an inverted output. The inputs, A and B, are **AND**ed, yielding AB, and then inverted, giving $\overline{AB}$. The symbol can be reduced from an **AND** gate followed by an inverter to the symbol shown in Fig. 4-34(b). The truth table shows that this logic function can be performed by using the hardware circuit illustrated in Fig. 4-33. Thus, the **NAND** gate can perform an **AND** function.

A second way of looking at the **NAND** circuit is to observe that its output is a 1 if A is 0 or (note that word) if B is 0. Therefore, it can perform the **OR** function. Figure 4-35 illustrates the symbol for this function. A is inverted, yielding $\bar{A}$ and B giving $\bar{B}$. Consequently, the **OR** gate forms $\bar{A} + \bar{B}$ at its output. A simplified form of the symbol is shown in Fig. 4-35(b). The circles should be interpreted as inverters placed

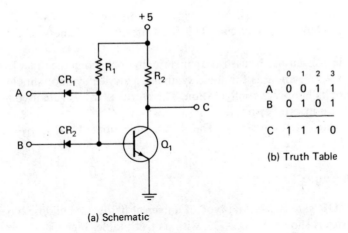

(a) Schematic

	0	1	2	3
A	0	0	1	1
B	0	1	0	1
C	1	1	1	0

(b) Truth Table

FIG. 4-33 NAND Gate Circuit

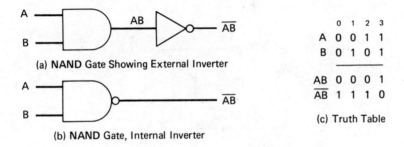

(a) **NAND** Gate Showing External Inverter

(b) **NAND** Gate, Internal Inverter

	0	1	2	3
A	0	0	1	1
B	0	1	0	1
AB	0	0	0	1
$\overline{AB}$	1	1	1	0

(c) Truth Table

FIG. 4-34 NAND Gate Used as **AND** Function

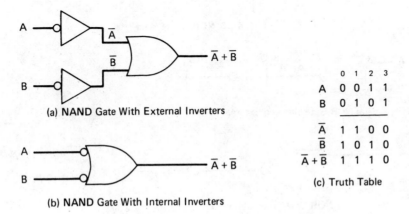

(a) **NAND** Gate With External Inverters

(b) **NAND** Gate With Internal Inverters

	0	1	2	3
A	0	0	1	1
B	0	1	0	1
$\overline{A}$	1	1	0	0
$\overline{B}$	1	0	1	0
$\overline{A} + \overline{B}$	1	1	1	0

(c) Truth Table

FIG. 4-35 NAND Gate Used as **OR** Function

in series with the input leads. Note that the truth table agrees with the truth table shown for the **NAND** gate circuit. This illustrates the usefulness of this unit as an **OR** gate.

This device can also be used as an inverter by connecting the input leads together. According to the standards for logic symbols, a device symbol should indicate its function, not its hardware configuration. Therefore, a **NAND** gate used as an inverter should have an inverter symbol.

One more interesting note. The demorganization of $\overline{AB}$ is $\bar{A} + \bar{B}$. They are equivalent by Law 21.

NOR Logic

The **NOR** gate is a second type of universal hardware building block. A circuit for this device is shown in Fig. 4-36, with its truth table. Figures 4-37 and 4-38 illus-

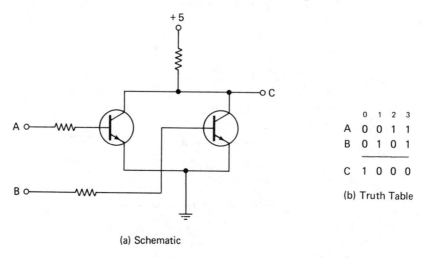

	0	1	2	3
A	0	0	1	1
B	0	1	0	1
C	1	0	0	0

(b) Truth Table

(a) Schematic

FIG. 4-36 NOR Gate Circuit

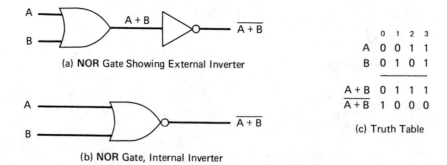

(a) **NOR** Gate Showing External Inverter

(b) **NOR** Gate, Internal Inverter

	0	1	2	3
A	0	0	1	1
B	0	1	0	1
A + B	0	1	1	1
$\overline{A + B}$	1	0	0	0

(c) Truth Table

FIG. 4-37 NOR Gate Used as **OR** Function

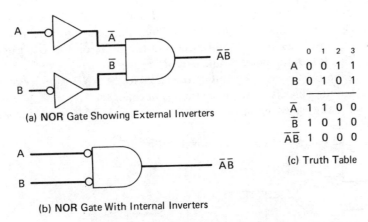

(a) NOR Gate Showing External Inverters

	0	1	2	3
A	0	0	1	1
B	0	1	0	1
$\overline{A}$	1	1	0	0
$\overline{B}$	1	0	1	0
$\overline{A}\overline{B}$	1	0	0	0

(c) Truth Table

(b) NOR Gate With Internal Inverters

FIG. 4-38 NOR Gate Used as **AND** Function

trate the logic symbols that can be used for the **NOR** gate. Both symbols provide truth tables whose outputs are identical to the circuit shown in Fig. 4-36. Thus, the **NOR** gate can perform both the **OR** function and the **AND** function. Note that demorganizing $\overline{A + B}$ gives $\overline{A}\overline{B}$, proving via Boolean algebra that both symbols are the same circuit.

As in the case of the **NAND** gate, the **NOR** can be used as an inverter by tying the input leads together.

Converting Circuits to Universal Logic

In most of the work in digital, circuits are computed and converted into **AND/OR**/invert logic. From there, the designer then selects either **NAND** or **NOR** logic for his system. He must subsequently convert his **AND/OR**/invert system to whichever universal system he has chosen. This can easily (there's that word again) be accomplished by the following procedure:

A. Draw the circuit in **AND/OR**/invert logic.
B. If **NAND** hardware has been chosen, add a circle to the outputs of each **AND** gate on the logic diagram, and provide circles on the inputs to all **OR** gates.
C. If **NOR** hardware has been chosen, add a circle to the output of each **OR** gate on the logic diagram, and provide circles on the inputs to all **AND** gates.
D. Add or subtract an inverter on each line that received a circle in step B or C.

Example 4-6: Convert Fig. 4-39(a) to **NAND** logic.

SOLUTION:

First, inverters are added on the inputs of each **OR** gate and the outputs of each **AND** gate.

Next, inverters are added to lines A and B so that the polarity on those lines remains unchanged from the original diagram. Observe that there are effectively two inverters (the inverter and the circle on the **OR** gate) on each line. This leaves the polarity unchanged.

Line U is inverted by adding the circle to the **AND** gate. By removing the inverter, the polarity of the line remains unchanged.

Finally, since F was inverted, an inverter is added, making its polarity unchanged.

The completed circuit is shown in Fig. 4-39(b).

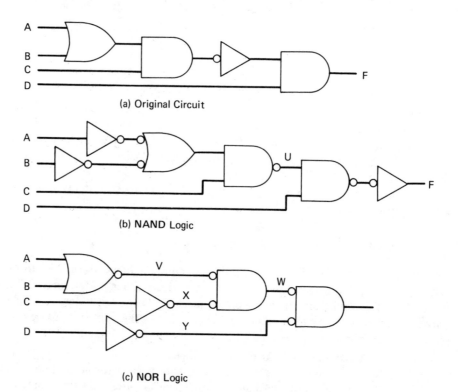

(a) Original Circuit

(b) **NAND** Logic

(c) **NOR** Logic

FIG. 4-39 Converting to Universal Logic

Conversion to **NOR** logic is accomplished by using a similar procedure.

Example 4-7: Convert Fig. 4-39(a) to **NOR** logic.

SOLUTION:

First, a circle is added to the output of each **OR** gate and to the inputs of each **AND** gate. Note that two circles have been added to line V. This is equivalent to a double inversion between the **OR** gate and the **AND** gate. Because of this, the polarity received by the **AND** gate is unchanged.

Inverters must be inserted on input leads C and D to cause the **OR** gate to receive the same polarity as the original circuit. The inverter in line W has been removed to compensate for the circle added to the **AND** gate.

The completed circuit is shown in Fig. 4-39(c).

A final word concerning **NAND** and **NOR** logic. From reading this section, you might conclude that the only hardware available is **NAND** and **NOR** gates. Although this was generally true in the 1960s, it is no longer true. **NAND, NOR, AND, OR,** and inverters are all off-the-shelf items today, allowing the designer maximum flexibility. Although he might use **NAND** logic as his primary element, he will probably supplement this with other types to reduce his integrated circuit package count.

4-12 NEGATIVE LOGIC

Thus far in this book logic levels have been defined as a 1 for $+5$ V and a 0 for ground or 0 V. This system is called positive logic because a 1 is more positive than a 0. In some systems it is convenient to define a ground level as a 1 and $+5$ V as a 0. This system is called *negative logic*.

Before going further, let us examine the term *assertion level*. In any logic system, something happens: a light turns on, an alarm rings, a letter is typed on a printer. These events happen because a signal is present on some logic line. If $+5$ V is required in order to "do something," we can say that the assertion level is $+5$ V. If a ground level is required to "do something," we can say the assertion level is negative. As you can see, the term *assertion level* refers to the level necessary to cause an event to occur.

Now, let us apply this concept of assertion level to positive and negative logic. In positive logic systems, positive assertion levels are defined by no bar over a function and negative assertion levels are defined by a bar over the function. In negative logic, negative assertion levels are defined by no bar over the function and positive assertion levels by a bar over the function.

It should be pointed out that negative logic changes only the definition of the equations. The logic building blocks with their circles remain unchanged. Figure 4-40 is a logic diagram using negative logic. When both switches are closed, the light turns on. The **AND** gate receives low true signals A and B. Rather than looking at the circles as inverters, look at them as assertion level indicators. Thus, this gate is one that will **AND** A (low) and B (low), giving a high (false) signal at its output. Its output is therefore $\overline{AB}$. In this system the lack of a circle indicates a positive assertion level and, therefore, a false logic level. $\overline{AB}$ is then inverted, becoming AB. When this level is grounded (true) the light illuminates.

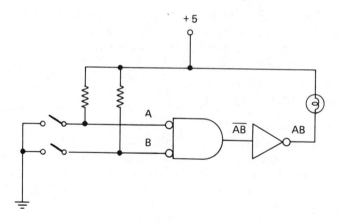

FIG. 4-40 Negative Logic

4-13 SUMMARY

Originating with rhetorical logic, Boolean algebra has been applied to the solution of digital logic problems. There are three fundamental building blocks in the system: the **AND** gate, the **OR** gate, and the inverter (the **NOT** function). Using these three operators and the rules which are summarized in Table 4-12, any logic expression can be reduced to its simplest form, thereby reducing the cost of its implementation.

TABLE 4-12. Boolean Algebra Laws

LAW 1	$\bar{0} = 1$
LAW 2	$\bar{1} = 0$
LAW 3	If $A = 0, \bar{A} = 1$
LAW 4	If $A = 1, \bar{A} = 0$
LAW 5	$\bar{\bar{A}} = A$
LAW 6	$A \cdot 0 = 0$
LAW 7	$A \cdot 1 = A$
LAW 8	$A \cdot A = A$
LAW 9	$A \cdot \bar{A} = 0$
LAW 10	$A + 0 = A$
LAW 11	$A + 1 = 1$
LAW 12	$A + A = A$
LAW 13	$A + \bar{A} = 1$
LAW 14	$A + B = B + A$
LAW 15	$A \cdot B = B \cdot A$
LAW 16	$A + (B + C) = (A + B) + C$
LAW 17	$A \cdot (B \cdot C) = (A \cdot B) \cdot C$
LAW 18	$A \cdot (B + C) = (A \cdot B) + (A \cdot C)$
LAW 19	$A + BC = (A + B)(A + C)$
LAW 20	$A + \bar{A}B = A + B$
LAW 21	$\overline{A\bar{B}} = \overline{A + B}$

NAND and NOR logic permit one item of hardware to perform the function of all three building blocks. Negative logic permits the designer to define all terms relative to negative assertion levels.

4-14 PROBLEMS

4-1. In what field of study did Boolean algebra originate?

4-2. What two men contributed the most to the development of Boolean algebra?

4-3. Name the three Boolean operators.

4-4. Give an example of a truth function using an **AND** function. An **OR** function. A **NOT** function.

4-5. Draw a Venn diagram to indicate all men who are bald but have beards.

4-6. Draw a Venn diagram indicating all planets that have moons that are not made of green cheese.

4-7. How many cases must be examined using the method of perfect induction for three variables? Five variables? Ten variables?

4-8. Prove that $AB + BC + C\bar{A} = AB + C\bar{A}$.

4-9. In the equation $T = A\bar{B}\bar{C} + AB$, indicate whether T is a 0 or a 1 under the following conditions:
   (a) $A = 1, B = 0, C = 1$ (b) $A = 0, B = 0, C = 0$
   (c) $A = 0, B = 1, C = 1$

4-10. In the equation $R = \overline{AB + C}$, indicate whether R is a 0 or a 1 under the following conditions:
   (a) $A, B,$ and $C = 0$ (b) $A = 1, B = 0, C = 0$
   (c) $A, B,$ and $C = 1$

4-11. Draw a truth table for the following equations:
   (a) $T = WX + XY$ (b) $V = R(\bar{S} + \bar{T})$

4-12. Draw a truth table for the following equations:
   (a) $M = N(P + R)$ (b) $M = N + P + N\bar{P}$

4-13. Draw a truth function table for crossing a river based upon whether the river is frozen over, the boat leaks, and whether you can swim.

4-14. Many cars produced in the United States have an interlock system that will allow the car to start only if both the front seat occupants have their seat belts on. Draw a truth function table indicating whether the car may be started based upon whether a passenger is present and whether both the passenger and the driver have buckled their seat belts.

4-15. Draw a truth function table expressing under what conditions I may put my trousers on, depending on whether I have my shorts on or off and my socks on or off. I must have my shorts on in order to don my trousers.

4-16. Repeat Problem 4-15 including whether or not I have my shoes on. I may not put on my shoes before I put on my trousers. Also, if I have dressed in the improper sequence, I must undress.

4-17. Reduce the following Boolean expressions:

    (a) $AABC$

    (b) $ABBBC$

    (c) $ABBCCC$

    (d) $A \cdot 1 \cdot A$

    (e) $A \cdot A \cdot A \cdot \bar{A}$

4-18. Reduce the following Boolean expressions:

    (a) $W + W + X$

    (b) $W + X + X + X$

    (c) $W + W + X + X + X$

    (d) $W + X + 1$

    (e) $W + X + 0$

4-19. Reduce the following Boolean expressions:

    (a) $AA + BB$

    (b) $A\bar{A} + BBB$

    (c) $BC + B\bar{B}C$

    (d) $WWX + WXX + WWX\bar{X}$

4-20. Reduce the following Boolean expressions:

    (a) $AA\bar{\bar{B}}B + \bar{A}BCC$

    (b) $A(\bar{A} + B)$

    (c) $X(WX + W\bar{X})$

    (d) $MNN(MN\bar{N} + MNP)$

4-21. Reduce the following Boolean functions:

    (a) $A + \bar{A}B + AB$

    (b) $A\bar{B} + \bar{A}B + AB + \bar{A}\bar{B}$

    (c) $A\bar{B}D + A\bar{B}D + \bar{B}D$

    (d) $W\bar{X}(W + Y) + WY(\bar{W} + \bar{X})$

    (e) $(AB + C)(AB + D)$

4-22. Reduce the following Boolean functions:

    (a) $A\bar{B}C + \bar{A}BC$

    (b) $A\bar{B}C + \bar{A}BC + ABC$

    (c) $(\bar{A}B)(AB) + AB$

    (d) $(1 + B)(ABC)$

    (e) $(A\bar{B} + A\bar{C})(BC + B\bar{C})(ABC)$

4-23. Reduce the following Boolean expressions:

    (a) $A + \bar{B}C(A + \overline{\bar{B}C})$

    (b) $C(\overline{ABC} + A\bar{B}C)$

    (c) $ABC + A\bar{\bar{B}} + BC$

    (d) $A[B + C(\overline{AB + AC})]$

4-24. Reduce the following Boolean expressions:

    (a) $\overline{AB + A(\overline{B + C})}$

    (b) $\overline{ABC(\overline{A + B + C})}$

    (c) $\overline{(\overline{AB} + ABC)(A\bar{B}C)}$

    (d) $(A + \overline{BC})(A\bar{B} + \overline{ABC})$

4-25. Without reducing, convert the following expressions to **AND/OR**/invert logic:

    (a) $AB + C(\bar{A} + B)$

    (b) $\overline{AB} + A + \overline{B + C}$

    (c) $A\bar{B} + \overline{AB}$

4-26. Without reducing, convert the following expressions to **AND/OR**/invert logic:

    (a) $A + B + AB$

    (b) $\overline{ABC} + A\bar{B}$

    (c) $(A + \bar{B} + C)(A\overline{BC})$

4-27. Without reducing, convert the expressions in Problem 4-21 to **NAND** logic.

4-28. Without reducing, convert the expressions in Problem 4-22 to **NOR** logic.

4-29. Reduce the expressions in Problem 4-21 to the simplest form and implement in **NAND** logic.

4-30. Reduce the expressions in Problem 4-22 to the simplest form and implement in **NOR** logic.

4-31. Reduce the expressions in Problem 4-21 to the simplest form and implement in **AND/OR**/invert logic.

4-32. Reduce the expressions in Problem 4-22 to the simplest form and implement in **AND/OR**/invert logic.

4-33. Assuming lines A and B are expressed in negative logic, draw the logic diagram for AB in **AND/OR**/invert logic.

4-34. Assuming lines A and B are expressed in negative logic, draw the logic diagram for $A + B$ in **AND/OR**/invert logic.

LOGIC
HARDWARE

5

5-1 THE PRACTICAL SIDE

So far within this book, **AND** gates, **OR** gates, and inverters have been described as ideal, elementary *black boxes* around which we developed elaborate design techniques. Don't you believe it! These black boxes are not ideal, for if they were, we would be missing out on half the fun. After all, boxes with purple spots, green stripes, and rounded corners are much more interesting than plain, old, black boxes.

The color and variety of these boxes come from the limitations of the elements used to construct logic elements. These transistor and integrated circuit limitations include rise and fall times, delay times, input and output voltage levels, noise immunity, and power dissipation. In this chapter we shall discuss these limitations, first presenting semiconductor switching phenomena. Since there are many excellent texts dealing with the analog uses of transistors and diodes, we shall only consider their application to switching circuits. The chapter will deal with the types of logic now available to the designer, discussing their parameters, limitations, advantages, and disadvantages.

5-2 THE DIODE AS A DC SWITCH

The semiconductor diode can be thought of as analogous to a simple switch. In the forward direction, the diode is ideally 0 ohms (Ω) of resistance, and in the reverse direction, infinite resistance. A switch has similar characteristics, 0 Ω when closed and infinite resistance when open. However, the diode is not an ideal device in either the forward or reverse direction.

In the forward direction, a typical silicon diode has the volt-amp characteristics shown in Fig. 5-1. Since the curve is not a straignt line, direct-current (dc) resistance varies, for V/I in one portion of the curve is not the same as V/I on another portion. If this were an ideal diode, its voltage drop would be 0 V at 5 milliamperes (mA), making its resistance $R = V/I = 0/5 = 0\ \Omega$. However, from the graph, the resistance is

$$R = \frac{V}{I} = \frac{0.8}{5.0} = 160\ \Omega$$

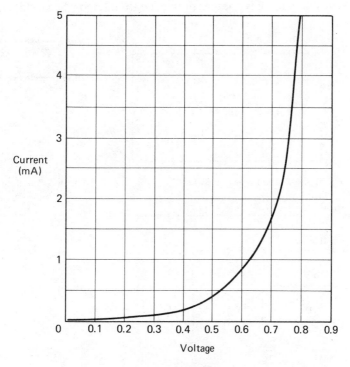

FIG. 5-1 Diode *VI* Curve

Although a forward resistance of 160 Ω does not seem substantial, consider its use in the circuit shown in Fig. 5-2, where this diode is to be used as a switch. The voltmeter will detect essentially 0 V when the diode is reverse biased. However, it will detect approximately 1 V when the diode is forward biased. If the diode were perfect, 2 V would be applied across the 160 Ω resistor. If the voltmeter were set to call any voltage above 1.2 V a 1 and any voltage less than 0.5 V a 0, the 1.0 V signal would be undefined.

This same reasoning can be applied to any other point on the curve. If, for example, the diode has a resistance of 1 kilohm (kΩ), a 1 kΩ load resistor in Fig. 5-2 could provide difficulties in defining a 1 or a 0.

Although its resistance is a problem, the voltage drop across a diode is even a greater source of inaccuracy. This voltage drop across a forward-biased silicon diode is typically 0.7 V, causing degeneration of the signal. Consider Fig. 5-3, a schematic of a diode **OR** gate. With + 5 V applied to the gate inputs, the output becomes 4.3 V. Note that 1's of + 5 V cause a 1 of 4.3 V. Extending this idea further (Fig. 5-4), 1's of 5 V at the input cause a 1 of 2.9 V out. Thus, voltage has degenerated from 5 V to 2.9 V. This can ultimately cause a 1 to be interpreted as a 0. The cure for this problem is to use an amplifier within each gate to build the signal back to its original, robust, healthy self.

Direct-current loading is another problem encountered when diodes are used as

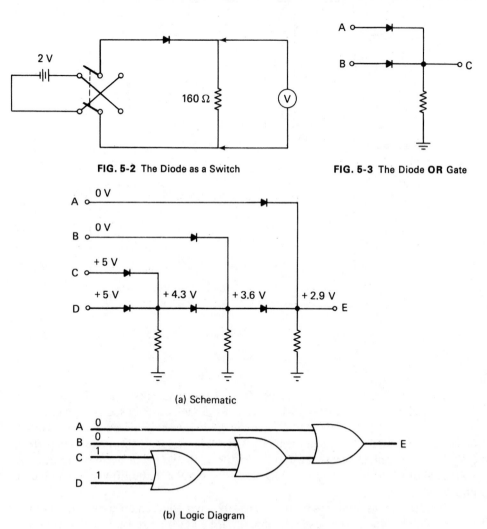

FIG. 5-2 The Diode as a Switch   **FIG. 5-3** The Diode **OR** Gate

(a) Schematic

(b) Logic Diagram

FIG. 5-4 Diode Voltage Drop Degeneration

gates. Consider the circuit shown in Fig. 5-5. With both inputs at + 5 V, the output, C, is + 5 V, as measured with a voltmeter. However, if a load, R_L, were connected to the output, the output voltage would be substantially reduced. For example, an R_L of 1 kΩ would provide an output voltage of 2.5 V, well below the desirable 5 V. And yet this is a very real problem, for many times an **AND** gate is connected to an **OR** gate, as shown in Fig. 5-6.

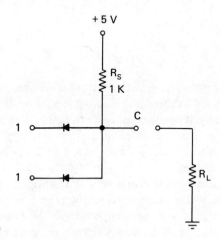

FIG. 5-5 Loaded Diode **AND** Gate

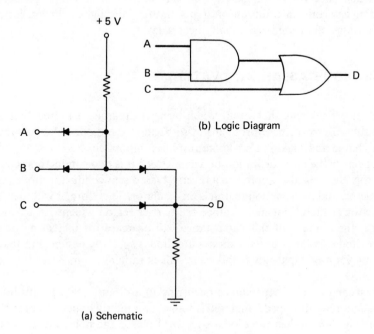

(b) Logic Diagram

(a) Schematic

FIG. 5-6 AND-OR Loading

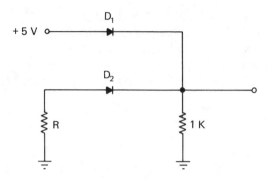

FIG. 5-7 Reverse Diode Leakage

Diodes are not perfect in the reverse direction, either—they leak (Fig. 5-7). If the gate were biased as shown, the output would be 4.3 V. If R were high and diode D_2 were quite leaky, it would be possible to build up a voltage drop across the resistor, and if this resistor were connected to another gate input, it might possibly supply a 1 to that circuit. Even a small voltage drop, 1 or 2 V, could cause problems if a noise spike were also present.

Power consumption is also a very real consideration within logic circuits. High power dissipation within resistors R_S and R_L (Fig. 5-5) requires large wires to minimize voltage drops, large power supplies to furnish the power, and cooling techniques to reduce the temperature of the diodes. However, using extremely low power levels for both true and false voltages makes the circuit susceptible to noise produced by the surrounding environment: fluorescent lights, motors, and light switches, for example. Therefore, some compromise is usually necessary.

5-3 THE DIODE AS AN AC SWITCH

There are two problems that occur when using a diode as a switch that affect its response time. Both of these problems arise because the diode exhibits capacitance, storing a charge and making immediate response impossible.

Remember the basic definition of a capacitor? It is two conductors separated by an insulator. But this also applies to a reverse-biased semiconductor diode; each lead is a conductor, and its depletion region is an insulator. Therefore, it will react much as any capacitor at high frequencies. Since the barrier region widens for an increase in dc voltage, the amount of this capacitance will decrease for increasing dc voltages across the diode. In fact, varactor diodes are diodes specially designed to take advantage of this varying capacitance and have values ranging from 3 to 100 picofarads (pF).

The second type of capacitance occurring in a diode is called diffusion capacitance. Assume a highly doped P material is joined to a lightly doped N material to form the diode (Fig. 5-8). When the diode is forward biased the holes will move to the N side in an attempt to unite with the free electrons. However, since there are few free

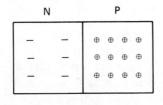

(a) Quiescent

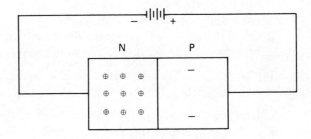

FIG. 5-8 Diffusion Capacitance   (b) Forward Biased

electrons with which to unite, many will be left standing at the church, never to know the pleasure of being married to an electron. On the other hand, the free electrons will move toward the *P* material in order to unite with a hole and, since there are great numbers of holes, will recombine. Therefore, we are left with many holes in the *N* material and very few carriers in the *P* material.

Next, observe what happens when the voltage is suddenly reversed. The free electrons, being more mobile than the holes (these little devils move twice as fast as the holes), move quickly to the *N* material. The slower-moving holes, therefore, leave a charge on the *N* side until they finally move to the *P* side. Therefore, we have current flow after the source has been shut off. This is the same as charging a capacitor, and then quickly shorting it out. There will be current flow after 0 V is applied across its terminals.

Both of these phenomena, junction capacitance and diffusion capacitance, prevent the diode from being an ideal switch.

5-4 THE BIPOLAR TRANSISTOR AS A DC SWITCH

Transistors can overcome the dc degeneration problem of the diode by their amplification. They make excellent switches in that they can be cut off to within a few megohms (MΩ) and can be turned on to a minimum resistance of a few ohms. Thus, they make very good logic elements. However, they are not ideal.

An ideal transistor would have the following characteristics:

A. Infinite input impedance:
   this is desirable in order that it will not load the driving stage and so that many transistors can be driven from the same driver.
B. Zero voltage drop from collector to emitter when saturated:
   this is required so that a logical 0 will be easily distinguished from a logical 1.
C. Infinite gain:
   this is desirable to prevent an input from loading a driver and to allow the driver to supply many gate inputs.
D. An output impedance of zero ohms:
   this is desirable so the output voltage will remain a constant, regardless of how many loads are connected to the transistor.

However, lest you be lulled into a false sense of security, note that transistors do not have these characteristics. Bipolar devices have, instead,

A. Low input impedance (1-5 kΩ resistance is typical in a common emitter).
B. A 50–100 millivolt (mV) emitter to collector voltage drop when saturated.
C. A current gain of 10–1000.
D. An output impedance of 10–200 kΩ. They are current generators rather than voltage generators.

Therefore, they are imperfect as logic elements. However, these problems are not severe enough to prevent their use. The most severe limitation is their low input impedance, necessitating much current flow and therefore much power dissipation. Metal oxide semiconductor devices overcome this limitation.

5-5 THE BIPOLAR TRANSISTOR AS AN AC SWITCH

The diode alternating-current (ac) problems discussed previously are also present in transistors. In the common emitter configuration, capacitance from the base to the collector provides negative feedback, preventing high frequencies from being properly amplified. This phenomenom, called the Miller effect, effectively amplifies the actual capacitance by the current gain of the transistor. Thus, even a small amount of base-collector capacitance will greatly affect the frequency response of the transistor. Since this capacitance depends on the area of the base-collector junction, high-frequency transistors are manufactured with smaller junctions.

Diffusion capacitance is also a problem in transistors. Since the emitter of a transistor is heavily doped to provide maximum emission and the base is lightly doped to prevent recombinations, reversing the bias of a forward-biased base-emitter junction leaves a diffusion charge in the base region that requires a finite time to discharge.

Therefore, saturating and then cutting off a transistor cannot be done instantaneously. This problem can be relieved by never allowing the transistor to go into saturation; emitter-coupled logic (ECL) is biased so that it is in the active region at all times and therefore has very small propagation delay times.

Effects Upon a Square Wave

This capacitance of transistors has a degenerating effect upon a square wave, for a perfect square wave is composed of all odd harmonics. Assume, for example, that the waveform in Fig. 5-9 is to be amplified and inverted by a bipolar transistor. If it were a perfect device, the waveform would appear as in Fig. 5-9(b). But, due to the capacitance within the transistor, its corners appear rounded [Fig. 5-9(c)]. Now rise time is defined as the time necessary for the waveform to rise from 10% of its maximum amplitude to 90% of its maximum. Therefore, instead of 0 picoseconds (ps) of rise time, the amplified waveform has a finite rise time T_r.

Let us also examine the delay time of the waveform, T_d. If we measure delay at 50% of the maximum height, the amplified waveform has also been delayed for a finite time. Therefore, this capacitance has two degenerating effects upon the wave form:

A. It increases rise time and
B. It delays the pulse.

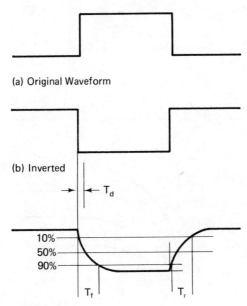

(a) Original Waveform

(b) Inverted

(c) Distortion

FIG. 5-9 Waveform Degeneration

5-6 THE FIELD EFFECT TRANSISTOR AS A SWITCH

Field effect transistors are being used in greater numbers as digital devices due to their extremely low power dissipation. The first effect transistors consisted of a bar of substrate material, type N, for example, and two P electrodes on each side [Fig.5-10(a)].

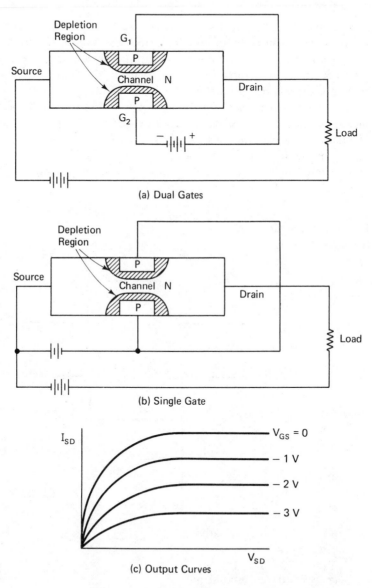

(a) Dual Gates

(b) Single Gate

(c) Output Curves

FIG. 5-10 The Junction Field Effect Transistor

When a voltage is applied between the source and drain, current flows through the N material. However, when a voltage is applied between G_1 and G_2 a depletion region forms around the gates, restricting N channel current flow to a narrow part of the channel where there are current carriers. Higher gate voltages result in a narrower channel until, if the voltage is high enough, N channel current is pinched off. Therefore, gate voltage controls drain current.

The next step was to tie both gates together and apply a voltage from the source to the gate [Fig. 5-10(b)]. This has the same effect, namely, allowing gate voltage to control drain current. This type of transistor is called a junction field effect transistor (JFET). These devices have very high input impedance, typically greater than 1 MΩ. Note that input voltage controls output current, whereas in the bipolar transistor, input current controls output current. A typical set of output curves is shown in Fig. 5-10(c), in the common source configuration. Note the constant current output with a constant voltage input.

Metal Oxide Semiconductors

The next stage of development was that of the metal oxide semiconductor field effect transistor (MOSFET). In this device, a low-resistivity N material is diffused onto a high-resistivity P material, called the substrate (Fig. 5-11). An insulator, silicon dioxide, is then formed by oxidizing the silicon substrate; the source and drain are etched through this insulator, forming electrodes for connection to the external circuit. The metallic gate electrode is deposited on the insulator but has no conductive connection to the substrate.

Applying a positive voltage to the drain results in a very small current flow from the source to the drain, since there are few minority carriers available. (Note that this is the same as applying a voltage from collector to emitter of a bipolar transistor with

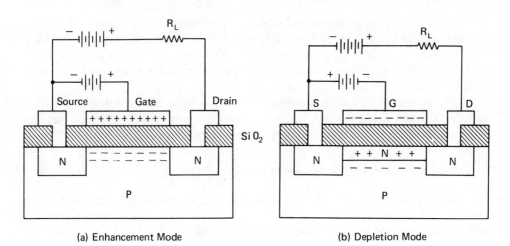

(a) Enhancement Mode   (b) Depletion Mode

FIG. 5-11 MOS Field Effect Transistors

the base open.) However, we shall now apply a positive voltage to the gate. This will attract many of the minority eletrons to the channel beneath the gate, providing more carriers for the source-drain current flow. With this new channel, electron flow is enhanced—hence the name enhancement mode.

A thin bar of N material can be added to the structure so that source-drain current will flow when the gate is at 0 V [Fig. 5-11(b)]. This current can be decreased by driving the gate negative, forcing the N-bar electrons into the P substrate and forming a depletion region around the gate. This type of device is called a depletion mode field effect transistor.

As can be seen, the input impedance of MOSFETs is extremely high, typically $10^{12}\,\Omega$, allowing many gate inputs to be driven from one output. In addition, its low source-drain voltage, typically less than 10 mV, and low power requirements, typically less than 5 microwatts (μW) quiescent, make it an excellent logic element. It does have one drawback, though. With those high input impedances, gate capacitances become quite a serious timing consideration. Remember that for a time constant $T_c = RC$. Therefore, high resistances, coupled with the Miller effect of gate to drain capacitance, increase circuit response times. However, industry is continually reducing these delay times through better fabrication techniques.

5-7 LOGIC SPECIFICATIONS

The specification sheets for the various logic families contain many terms unique to digital electronics. Before detailing the characteristics of each family, let us discuss several of the specifications we are about to encounter.

Fan-in and Fan-out

The fan-in of a logic circuit is the number of inputs that logic circuit can handle. For example, an eight-input gate requires one unit load per input. Therefore, its fan-in is 8. If it were expandable by use of an expander (this is an element with multiple inputs designed to be connected to a gate to increase the number of inputs that gate can handle), then its fan-in would be greater than 8.

The fan-out of a circuit is the number of unit inputs that can be driven by a logic element. If a gate has a fan-out of 6, this means it can drive six unit inputs and still maintain its logical 1 and logical 0 output voltage specifications.

Noise Immunity and Noise Margin

Noise immunity represents the amount of noise a logic system can handle without being amplified beyond unity again. If, for example, 20 inverters were connected in series, the amount of ac noise immunity would be that which if a spike of that voltage were to appear at the input would cause a spike of identical amplitude to appear at the output. A logic inverter is an amplifier. However, unless the input reaches a par-

ticular amplitude, the inverter remains in saturation, the output unchanged. Therefore, at low levels of input spikes, the inverter has a gain of 0. At very high levels of
input, it may have a gain of 5. For example, assume 1 V or greater to a logic family
produces a logical 1 on the input, causing a 5 V logic swing on the output, giving a
gain of 5 (Fig. 5-12). Therefore, there is some spike amplitude at which the gain is 1.
This amplitude is called the device's noise immunity, V_{NI}.

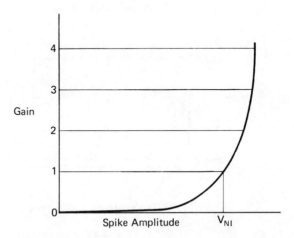

FIG. 5-12 Noise Immunity

A second noise specification is that of noise margin. Assume a circuit consists of
two inverters (Fig. 5-13). According to the manufacturer, the highest logical 0 any
inverter will ever output is 0.2 V. However, the manufacturer also guarantees that
any input less than 1.0 V will always be considered a logical 0. Therefore, we can
have $1.0 - 0.2 = 0.8$ V of noise on the input to the second stage and still call it
a logical 0. This 0.8 V "margin of safety" is called the noise margin, V_{NM} (Fig. 5-13).
Stated mathematically,

$$V_{NML} = V_{ILmax} - V_{OLmax}$$

The same system of reasoning applies to high-level noise margin, V_{NMH}. If the manufacturer guarantees the minimum logical 1 at a gate output to be 4.0 V but states that
any voltage down to 3.6 V is to be considered a logical 1 input, the noise margin is

$$V_{NMH} = V_{OHmin} - V_{IHmin}$$
$$= 4.0 - 3.6$$
$$= 0.4 \text{ V}$$

Propagation Delay

Propagation delay is the time it takes a pulse to get through a logic device, measured at 50% of peak (Fig. 5-14). This is an important consideration, for it determines how fast the logic system can operate.

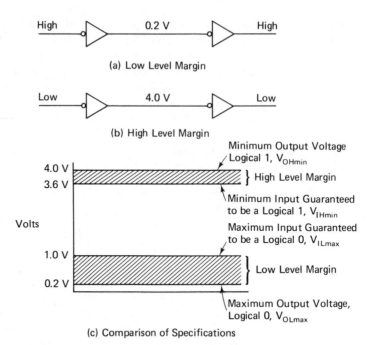

(a) Low Level Margin

(b) High Level Margin

(c) Comparison of Specifications

FIG. 5-13 Noise Margins

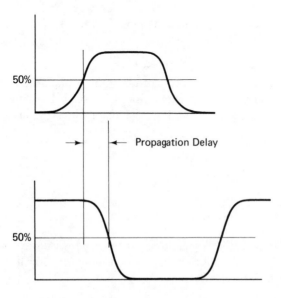

FIG. 5-14 Propagation Delay

5-8 LOGIC FAMILIES

A number of logic families are on the market today, and variations are introduced continually. Therefore, the major families will be discussed, along with their advantages and disavantages.

Resistor-Transistor Logic (RTL)

The RTL family was the first to be introduced. An RTL **NOR** gate is shown in Fig. 5-15. If input A is high, Q_1 will conduct, pulling pin C, the output, to ground.

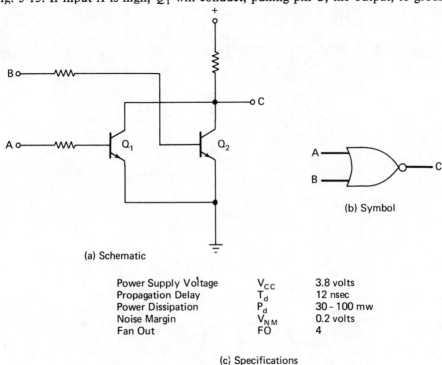

(a) Schematic

(b) Symbol

Power Supply Voltage	V_{CC}	3.8 volts
Propagation Delay	T_d	12 nsec
Power Dissipation	P_d	30 - 100 mw
Noise Margin	V_{NM}	0.2 volts
Fan Out	FO	4

(c) Specifications

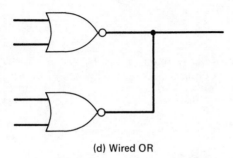

(d) Wired OR

FIG. 5-15 Resistor-Transistor Logic (RTL)

117

Similarly, if B is high, C will be pulled to ground through Q_2. The specifications are summarized in Fig. 5-15(c).

Both RTL and DTL have a capability called wire **OR** [Fig. 5-15(d)]. Since the output is effectively a transistor, two outputs can be wired together, producing a low true **OR** function. This saves on the total packages needed for the design.

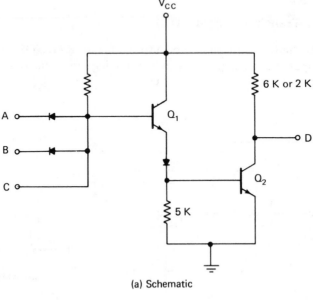

(a) Schematic

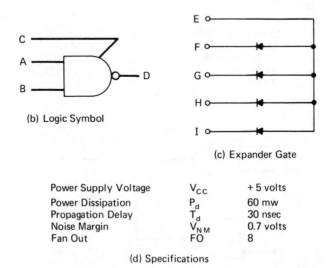

(b) Logic Symbol

(c) Expander Gate

Power Supply Voltage	V_{CC}	+ 5 volts
Power Dissipation	P_d	60 mw
Propagation Delay	T_d	30 nsec
Noise Margin	V_{NM}	0.7 volts
Fan Out	FO	8

(d) Specifications

FIG. 5-16 Diode-Transistor Logic (DTL)

Diode-Transistor Logic (DTL)

The next family that was introduced was DTL. A two-input DTL **NAND** gate is shown in Fig. 5-16. If lead A is grounded, transistor Q_1 is cut off, preventing any base current from being provided Q_2. Therefore, Q_2 will cut off, leaving point D high. Lead B will act in the same manner.

Lead C is called an expander input. If lead E of Fig. 5-16(c) is connected to this lead, the gate will become a six-input gate. This allows a great deal of versatility in the design.

DTL can be purchased with either 6 KΩ or 2 KΩ load resistors. This allows the designer to vary his power dissipation. However, lower power dissipation using the 6 KΩ resistor also introduces greater propagation delays. (Remember, $T_c = RC$.)

Transistor-Transistor Logic (TTL)

Also known as T^2L, this family was introduced to provide greater speed than DTL. A 74 Series TTL **NAND** gate is shown in Fig. 5-17, with its logic symbol and specifications. If both A and B are high, Q_1 has no emitter current; however, its base-collector junction is forward biased, supplying base current to Q_2. This, in turn, feeds base current to Q_4, causing it to conduct. However, the Q_2 collector will go low, cutting off Q_3. Therefore, we have Q_4 conducting and Q_3 cut off.

If either A or B goes low, Q_1 will have base-emitter current, saturating it and pulling Q_2 base to ground, cutting it off. This will cause Q_3 to conduct and Q_4 to cut off. Note that this is a logic family that conducts when on or off. This means that the output line will always have a low impedance, reducing the effect of noise.

TTL cannot be wire **OR**ed. However, TTL gates are available with Q_3 missing. This enables a designer to supply his or her own resistor to V_{CC}, and allows the **OR** gates to be wired.

Unused inputs on TTL gates are subject to noise and should be connected to a defined logic level. They may be connected to either a used input or through a resistor to V_{CC}. These resistors are called *pull-up* resistors and make sure the base-emitter junction of the input transistor is reverse biased.

TTL has several important subfamilies. The 74LS (low-powered Schottky) series has both higher speed and lower power dissipation than the 74 series (Fig. 5-18). Diode D_1 and others of this figure are called Schottky diodes after their inventor. These diodes have very low saturation voltages, on the order of 0.4 V, compared with 0.6 V for diffused diodes. When Schottky diodes are connected as shown in Fig. 5-19, they form a Schottky transistor. During saturation of a silicon transistor, the base is at 0.7 V and the collector at 0.1 V with respect to the emitter. By placing the Schottky diode as shown, the collector cannot be driven more than 0.4 V below the base. Thus, the transistor is effectively prevented from going into saturation. This reduces diffusion capacitance and propagation delay.

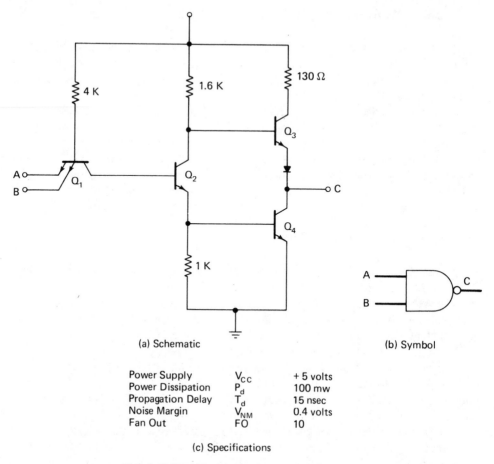

(a) Schematic   (b) Symbol

Power Supply	V_{CC}	+ 5 volts
Power Dissipation	P_d	100 mw
Propagation Delay	T_d	15 nsec
Noise Margin	V_{NM}	0.4 volts
Fan Out	FO	10

(c) Specifications

FIG. 5-17 74 Series Transistor-Transistor Logic (TTL)

The 74S series TTL is even faster than the 74LS series: 5 ns versus 10 ns. How-ever, a penalty of higher dissipation is paid for this feature: 180 mw versus 22 mw. Its circuit is similar to that of Fig. 5-18, except that it has lower resistor values and an input of multiple emitters as in Q_1 of Fig. 5-17.

It should be noted that, although all TTL subfamilies are compatible, their input and output currents differ.

Integrated Injection Logic (I²L)

In this system, Fig. 5-20, Q_1 and Q_3 act as current sources to the bases of Q_2 and Q_4, respectively. If input A goes low, the current to the base of Q_2 will be shorted to ground, preventing this transistor from conducting. In a similar manner, input B controls Q_4. Thus, if either point A or point B is high, the output will be low, a **NOR** function.

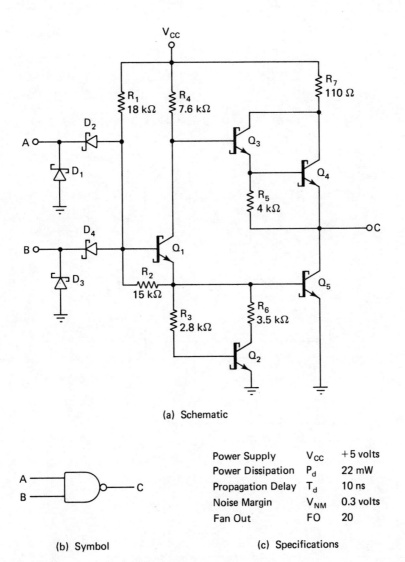

(a) Schematic

A ——⌐‾‾‾⌐
B ——⌐___⌐⊸— C

(b) Symbol

Power Supply	V_{CC}	+5 volts
Power Dissipation	P_d	22 mW
Propagation Delay	T_d	10 ns
Noise Margin	V_{NM}	0.3 volts
Fan Out	FO	20

(c) Specifications

FIG. 5-18 74LS Series TTL

The big advantage of I²L over similar bipolar logic can be found by observing that the collector of Q_1 is made of the same material as that of the base of Q_2, making it possible to be the same physical element, composed of P type material. This greatly improves the density that a given chip can contain. Note further that there are no resistors in the circuit, further reducing the size of the gate. Therefore, I²L is used in large scale functions where bipolar devices are required. It is also slightly faster (5 ns) than TTL.

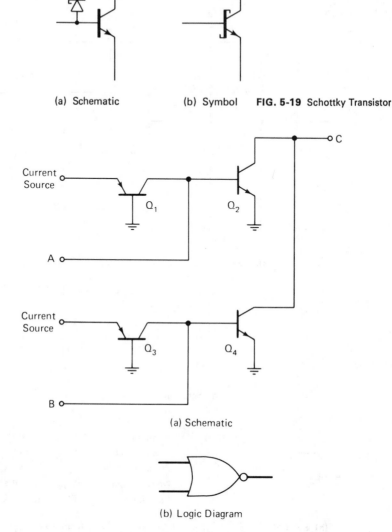

(a) Schematic   (b) Symbol   **FIG. 5-19** Schottky Transistor

(a) Schematic

(b) Logic Diagram

FIG. 5-20 Integrated Injection Logic (I²L)

Complementary Metal Oxide Semiconductor (CMOS) Logic

Although there are both MOS and CMOS families on the market, we shall concentrate on CMOS devices. Figure 5-21 shows a schematic for a two-input **NOR** gate. With A and B high, Q_1 and Q_2 are cut off and Q_3 and Q_4 are turned on, causing C to go to ground. When A goes to ground, Q_1 will conduct and Q_3 will turn off, causing C to go high. Similarly, when B goes to ground, Q_4 will cut off and Q_2 turn on, causing C to go high. Therefore, like TTL, one output transistor is always conducting.

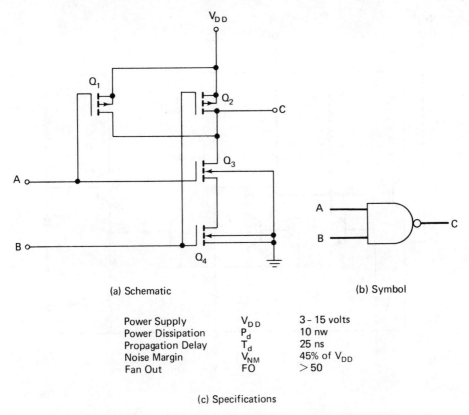

(a) Schematic

(b) Symbol

Power Supply	V_{DD}	3 - 15 volts
Power Dissipation	P_d	10 nw
Propagation Delay	T_d	25 ns
Noise Margin	V_{NM}	45% of V_{DD}
Fan Out	FO	> 50

(c) Specifications

FIG. 5-21 Complementary Metal Oxide Semiconductor Logic (CMOS)

Unused inputs should always be connected to a defined logic level; otherwise, its high input impedance would transmit any noise pulses that come along.

Emitter-Coupled Logic (ECL)

Emitter-coupled logic is designed for speed. It is the only family that does not operate fully saturated or cut off, preventing the adverse effects of diffusion capacitance. Output levels are -0.8 V for a logical 1 and -1.8 V for a logical 0. A schematic, logic symbol, and list of specifications are shown in Fig. 5-22.

Causing A to go high will drive the Q_2 collector low, causing both the base and emitter of Q_1 to go low, outputting a low on the **NOR** lead. At the same time, the Q_4 emitter will go high, tending to cut it off, and the Q_5 base will go high, as will the Q_5 emitter, and, consequently, the **OR** output. D_1 and D_2 provide a constant voltage supply to the Q_6 base and, therefore, a constant voltage to the Q_4 base so that it can operate as a common base amplifier.

5-9 LOGIC PACKAGES

Logic elements can be purchased in a variety of packages, including dual-in-line package (DIP), flat packs, and TO-5 shapes (Fig. 5-23). The DIP is the most widely used

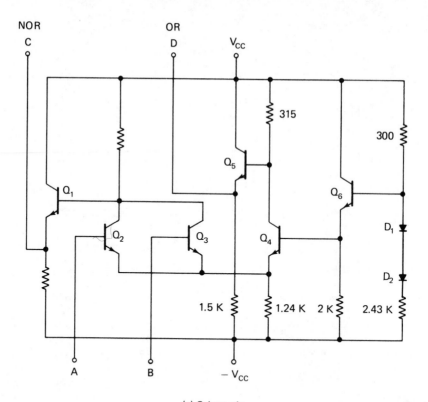

(a) Schematic

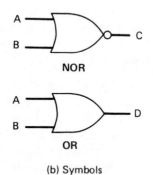

(b) Symbols

Power Supply	V_{CC}	− 5.2 volts
Power Dissipation	P_d	100 mw/pkg
Propagation Delay	T_d	1 ns
Fan Out	FO	16

(c) Specifications

FIG. 5-22 Emitter Coupled Logic (ECL)
(Courtesy Texas Instruments, Incorporated)

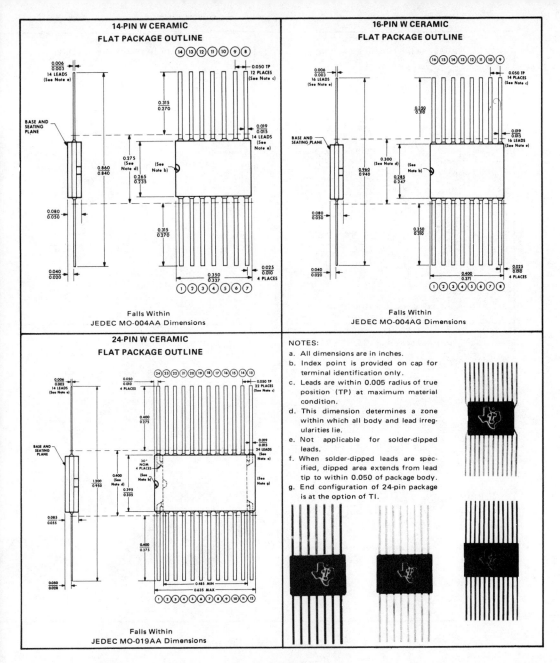

W ceramic flat packages

These hermetically sealed flat packages consist of an electrically nonconductive ceramic base and cap, and a 14-, 16-, or 24-lead frame. Hermetic sealing is accomplished with glass. Tin-plated ("bright-dipped") leads (—00) require no additional cleaning or processing when used in soldered assembly.

(a)

FIG. 5-23 Package Outlines
Courtesy Texas Instruments, Incorporated)

125

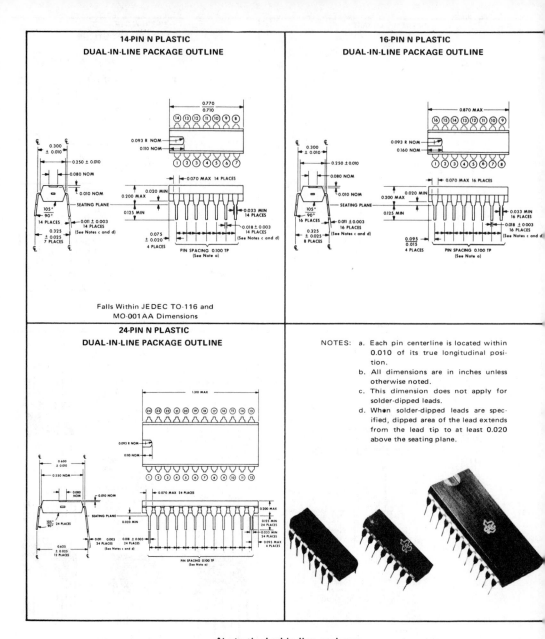

N plastic dual-in-line packages

These dual-in-line packages consist of a circuit mounted on a 14-, 16-, or 24-lead frame and encapsulated within an electrically nonconductive plastic compound. The compound will withstand soldering temperature with no deformation and circuit performance characteristics remain stable when operated in high-humidity conditions. These packages are intended for insertion in mounting-hole rows on 0.300-inch (or 0.600-inch) centers. Once the leads are compressed and inserted, sufficient tension is provided to secure the package in the board during soldering. Silver-plated leads (−00) require no additional cleaning or processing when used in soldered assembly.

(b)

FIG. 5-23 (continued)

because it can be easily inserted into either an integrated circuit (IC) socket or holes in a printed circuit card. Both DIP and flat packs are available in 8-pin to 48-pin ceramic or plastic packages to accommodate a wide variety of complex circuits. The ceramic packages are designed for severe environments, whereas the plastic packages are designed for economy.

Another very important specification of the package is its range of operating temperatures. Although manufacturers differ, the following choices are usually available:

a. Military range $-55°$ to $+ 125°C$
b. Quasi-military range $-40°$ to $+ 80°C$
c. Industrial range $0°$ to $+ 70°C$

5-10 SUMMARY

Although diodes can be used as gates, they cause logic level degeneration. Therefore, both bipolar and field effect transistors are used as a part of logic gates to prevent this degeneration. However, they are imperfect, providing delays to digital pulses. Several logic families are available to the designer: RTL, DTL, TTL, MOS, and ECL. Each has advantages and disavantages, requiring the designer to choose which specifications are most important to a particular project. Additionally, the designer must select the package and temperature ranges required for the job.

5-11 PROBLEMS

5-1. Name two dc problems encountered when using a diode as a gate element.

5-2. Name and describe two types of diode capacitances.

5-3. What limitation of a bipolar transistor does a field effect transistor overcome?

5-4. Describe the Miller effect.

5-5. At what points on a waveform is a rise time measured?

5-6. What disadvantage does the field effect transistor have that the bipolar transistor does not?

5-7. Name the two operating modes of a MOSFET.

5-8. A four-input gate can drive 10 similar gate inputs. What is its fan-in? Its fan-out?

5-9. A certain logic family specifies that the highest logical 0 output of a gate is 2.0 V. The gate input will be interpreted as a 0 for any voltage below 2.8 V. In addition, a 3 V spike applied to the gate input will appear on the output of the gate as a 3 V spike. What is the noise immunity of this family? The noise margin?

5-10. Propagation delay is measured at what point on the waveform?

5-11. What was the first logic family?

5-12. Which logic families can always be wire-ORed?

5-13. Which logic families always have one output transistor conducting?

5-14. What logic family has the smallest propagation delay?

5-15. What IC package shape is most popular?

5-16. Name two materials used for IC packages.

COMBINATIONAL LOGIC

6

6-1 INTRODUCTION

In this chapter we shall discuss combinational logic, a system in which the output occurs in direct, immediate response to input stimuli. This is the type of logic we have come to know and love from the previous chapters within this text. Figure 6-1 is a block diagram of such a combinational logic system. A 1 appears at the output every time a certain combination appears at the inputs. An example of this would be a four-input **NAND** gate that, when all inputs are high, immediately produces a low at the output in response.

There are two useful techniques for reducing combinational logic equations and logic diagrams to the fewest possible elements: mapping and tabular minimization. Mapping, discussed first, is the "workhorse" of these techniques. Tabular minimization, discussed in the latter part of this chapter, is more difficult, more accurate, and applicable as a computerized reduction technique. Little did you suspect that when you picked up this book you would prepare yourself to be an expert cartographer.

In contrast to combinational logic, sequential logic introduces the element of time into the system. Not only must the right combinations appear at the inputs, but they must appear in the right order with respect to time, similar to a combination on a safe.

6-2 MINTERMS

One of the most powerful theorems within Boolean algebra states that any Boolean function can be expressed as the sum (**OR**) of products (**AND**) of all the variables within the system. For example, $A + B$ can be expressed as the sum of several prod-

FIG. 6-1 Combinational Logic Block Diagram

ucts, each of which contains the letters A and B. These products are called minterms, and each contains all the letters within the system, either with or without the bar.

Example 6-1: Convert $A + B$ to minterms.

SOLUTION:

$$A + B = A(1) + B(1)$$
$$= A(B + \bar{B}) + B(A + \bar{A})$$
$$= AB + A\bar{B} + AB + \bar{A}B$$
$$= AB + A\bar{B} + \bar{A}B$$

Note that each term in the example contains all the letters used: A and B. The terms AB, $A\bar{B}$, and $\bar{A}B$ are therefore minterms. This process of "unreducing" a Boolean expression is called *expansion* and is carried out using the following procedure:

A. Write down all the terms.
B. Put X's where letters must be provided to convert the term to a minterm.
C. Use all combinations of the X's in each term to generate minterms. Where an X is a 0, write a barred letter; where it is a 1, write an unbarred letter.
D. Drop out redundant terms.

Example 6-2: Find the minterms for $A + BC$.

SOLUTION:

Write down terms	$A + BC$
Insert X's where letters are missing	$AXX\ XBC$
Vary all the X's in AXX	$A\bar{B}\bar{C},\ A\bar{B}C,\ AB\bar{C},\ ABC$
Vary all the X's in XBC	$\bar{A}BC,\ ABC$

Therefore,

$$A + BC = A\bar{B}\bar{C} + A\bar{B}C + AB\bar{C} + ABC + \bar{A}BC + ABC$$
$$= A\bar{B}\bar{C} + A\bar{B}C + AB\bar{C} + ABC + \bar{A}BC$$

Just to make it interesting, prove that the result is $A + BC$.

$$A\bar{B}\bar{C} + A\bar{B}C + AB\bar{C} + \bar{A}BC + ABC$$
$$= A\bar{B}(C + \bar{C}) + AB\bar{C} + BC(\bar{A} + A)$$
$$= A\bar{B} + AB\bar{C} + BC$$
$$= A\bar{B} + B(A\bar{C} + C)$$

$$= A\bar{B} + B(A + C)$$
$$= A\bar{B} + AB + BC$$
$$= A(\bar{B} + B) + BC$$
$$= A + BC$$

Example 6-3: Find the minterms for $AB + ACD$.

SOLUTION:

$ABXX$ generates $AB\bar{C}\bar{D}, AB\bar{C}D, ABC\bar{D}, ABCD$

$AXCD$ generates $A\bar{B}CD, ABCD$

Therefore,

$$AB + ACD = AB\bar{C}\bar{D} + AB\bar{C}D + ABC\bar{D} + A\bar{B}CD + ABCD$$

Since the terms A, B, C, and D must appear in every product, a shorthand notation has been developed that saves actually writing down the letters themselves. To form this notation, develop a binary word out of the letters, substituting a 0 for a **NOT**ed letter and a 1 for a letter without a bar. Then express the decimal equivalent of this binary word as a subscript of a lowercase m.

Example 6-4: Find the minterm designation of $A\bar{B}\bar{C}\bar{D}$.

SOLUTION:

Copy original term	$A\bar{B}\bar{C}\bar{D}$
Substitute 1's for nonbarred letters and 0's for barred letters	$1\,0\,0\,0$
Express as decimal subscript of m	m_8

Therefore, $A\bar{B}\bar{C}\bar{D} = m_8$.

Example 6-5: Find the minterm designation of $\bar{W}\bar{X}Y\bar{Z}$.

SOLUTION:

Original term	$\bar{W}\bar{X}Y\bar{Z}$
Convert to binary	$0\,0\,1\,0$
Express as decimal subscript of m	m_2

Therefore, $\bar{W}\bar{X}Y\bar{Z} = m_2$.

It should be emphasized that this subscript will be used in minterm designators (m_8, m_{13}, etc.), truth tables, and maps.

6-3 TRUTH TABLES AND MAPS

Although the value of truth tables was demonstrated in Chapter 4, there was no attempt made to identify and number each column of the table. There are several configurations of tables used by different design organizations; however, only one will be presented, and that one will have letters and numbers consistent with maps and terms used throughout the rest of this book. Therefore, although this chapter does not have the only right answer, it has a consistent set of right answers.

Two Variables

Table 6-1 illustrates a truth table for two variables, A and B. The numbers above the columns are minterm designators, m_0, m_1, m_2, and m_3. Minterm 0 occurs when the A is 0 (**NOT**ed) and the B is 0; therefore, m_0 represents $\bar{A}\bar{B}$. Note that, to be consistent, the A will always represent the most significant bit of the designator. Minterm 1, m_1, occurs when A is 0 and B is 1; it therefore represents $\bar{A}B$. Minterm 3, m_3, occurs when both A and B are 1's, as in AB.

TABLE 6-1. Two-Variable Truth Table

	0	1	2	3
A	0	0	1	1
B	0	1	0	1
C	0	1	1	1

As can be seen, each column of the truth table represents a minterm. Below the table, a line is drawn to indicate any output terms. Below the line, 1's are placed to indicate that the output contains a particular minterm in its sum and 0's when that term is excluded from the sum. According to Table 6-1, the output, C, contains minterms 1, 2, and 3. That is,

$$C = \bar{A}B + A\bar{B} + AB$$

Another way of representing this relationship is to use the Greek letter sigma and an m to represent "the sum of minterms. . . ." Applying this notation to Table 6-1,

$$C = \sum m(1, 2, 3)$$

This should be read "C is the sum of minterms 1, 2, and 3." As the broken record says, this type of notation is easy to learn and easy to manipulate.

The two-variable map (Fig. 6-2) is used with the two-variable truth table and represents essentially the same information. Each square represents one unique minterm, as shown by the small Arabic figure in the square. To define an output term, a large 1 is placed in the square for minterms to be included and 0's in those squares to be excluded. Accordingly, Fig. 6-2 reads

$$C = \sum m(1, 2, 3)$$

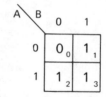

FIG. 6-2 Two-Variable Map

Example 6-6: Chart on a truth table and map $C = \bar{A}\bar{B} + A\bar{B}$.

SOLUTION:

$\bar{A}\bar{B}$ is m_0, and $A\bar{B}$ is m_2.
Charting on a truth table,

$$
\begin{array}{c c c c c}
 & 0 & 1 & 2 & 3 \\
A & 0 & 0 & 1 & 1 \\
B & 0 & 1 & 0 & 1 \\
\hline
C & 1 & 0 & 1 & 0 \\
\end{array}
$$

Mapping,

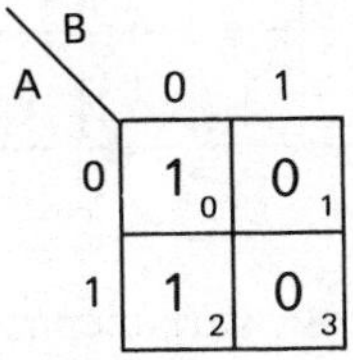

Three Variables

Figure 6-3 illustrates a three-variable map and its associated truth table. The binary numbers along the top of the map indicate the condition of B and C for each column. For example, 10 indicates $B\bar{C}$ is part of the product, as in minterm 6 when A is 1 and BC is 10, making the whole term $AB\bar{C}$.

The truth table has been expanded to include the eight states necessary for three variables. Again, note that A is the most significant bit of the minterm designator and C the least significant bit. Output terms are indicated in the same manner as on the two-variable map.

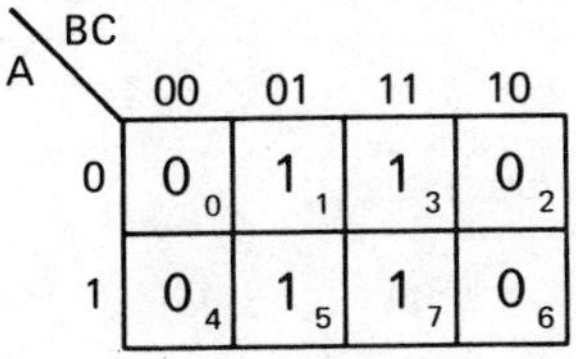

	0	1	2	3		4	5	6	7
A	0	0	0	0		1	1	1	1
B	0	0	1	1		0	0	1	1
C	0	1	0	1		0	1	0	1
D	0	1	0	1		C	1	0	1

$D = \Sigma\, m(1, 3, 5, 7)$

$D = \bar{A}\bar{B}C + \bar{A}BC + A\bar{B}C + ABC$

(a) Map   (b) Truth Table   (c) Equations

FIG. 6-3 Three-Variable Map

Example 6-7: Chart and map the equation $X = ABC + A\bar{B}C + A\bar{B}\bar{C}$.

SOLUTION:

$ABC = m_7$, $A\bar{B}C = m_5$, and $A\bar{B}\bar{C} = m_4$.

Charting,

```
            0  1  2  3     4  5  6  7
   A        0  0  0  0     1  1  1  1
   B        0  0  1  1     0  0  1  1
   C        0  1  0  1     0  1  0  1
           ─────────────────────────
   X        0  0  0  0     1  1  0  1
```

Mapping,

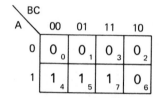

Four Variables

Figure 6-4 illustrates the four-variable map and truth table. Since 2^4 equals 16, there are 16 unique states labeled 0 through 15. The binary numbers along the top of the map indicate the trueness or **NOT**ness of the CD portion of the minterm and those along the sides the AB portion. Using this scheme, $\bar{A}B\bar{C}\bar{D}$ would be in the $\bar{A}B$ horizontal row (binary 01) and the $\bar{C}\bar{D}$ vertical column (binary 00). Since 0100 is a binary 4, $\bar{A}B\bar{C}\bar{D}$ represents minterm 4, as shown by the small 4 on the map.

The truth table is an expansion of the three-variable truth table so that 16 states may be accommodated.

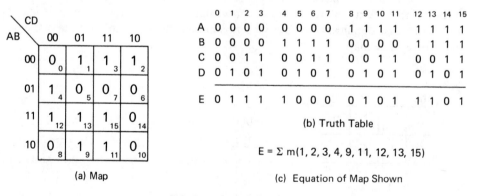

(a) Map

(b) Truth Table

$$E = \Sigma\, m(1, 2, 3, 4, 9, 11, 12, 13, 15)$$

(c) Equation of Map Shown

FIG. 6-4 Four-Variable Map

Five Variables

Figure 6-5 illustrates a five-variable map. Because 32 terms are required, the map has been divided into two 16-square blocks. The left block represents all those minterms in which A is a 0 (minterms 0 through 15). The right block represents all

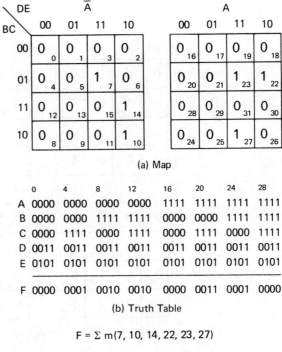

(a) Map

	0	4	8	12	16	20	24	28
A	0000	0000	0000	0000	1111	1111	1111	1111
B	0000	0000	1111	1111	0000	0000	1111	1111
C	0000	1111	0000	1111	0000	1111	0000	1111
D	0011	0011	0011	0011	0011	0011	0011	0011
E	0101	0101	0101	0101	0101	0101	0101	0101
F	0000	0001	0010	0010	0000	0011	0001	0000

(b) Truth Table

$$F = \Sigma\, m(7, 10, 14, 22, 23, 27)$$

FIG. 6-5 Five-Variable Map   (c) Equation for Above Map

those minterms in which A is a 1 (minterms 16 through 31). Other than these differences, the map and table are the same as for four variables.

Six Variables

A six-variable map is shown in Fig. 6-6. The left two blocks represent terms in which B is 0, and the right two blocks, terms in which B is 1. Similarly, the top two blocks represent terms in which A is 0, and the bottom two, terms in which A is 1. These blocks have the same arrangement as a two-variable map of A and B.

Practical Mapping Procedures

Although the above maps and truth tables can be used as shown, there are three shortcuts that will reduce the amount of writing required. Since the pattern of the square designation is so regular, most designers omit numbering each square and count up from 0, 16, 32, or 48, starting from the upper left. Similarly, most omit the minterm designation numbers from the truth tables. Another shortcut that can be used is to show a large 0 to represent 0000 in a truth table and a large 1 to represent 1111. Figure 6-7 illustrates these three procedures as applied to a four-variable map.

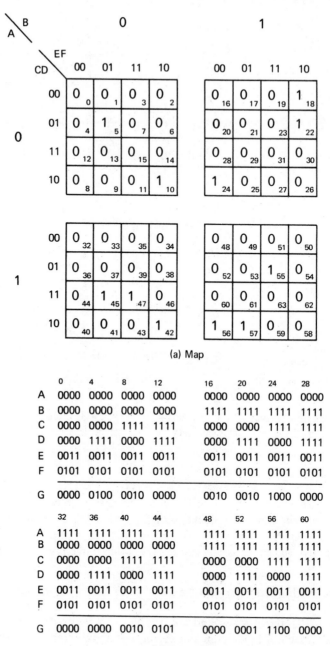

(a) Map

	0	4	8	12		16	20	24	28
A	0000	0000	0000	0000		0000	0000	0000	0000
B	0000	0000	0000	0000		1111	1111	1111	1111
C	0000	0000	1111	1111		0000	0000	1111	1111
D	0000	1111	0000	1111		0000	1111	0000	1111
E	0011	0011	0011	0011		0011	0011	0011	0011
F	0101	0101	0101	0101		0101	0101	0101	0101
G	0000	0100	0010	0000		0010	0010	1000	0000

	32	36	40	44		48	52	56	60
A	1111	1111	1111	1111		1111	1111	1111	1111
B	0000	0000	0000	0000		1111	1111	1111	1111
C	0000	0000	1111	1111		0000	0000	1111	1111
D	0000	1111	0000	1111		0000	1111	0000	1111
E	0011	0011	0011	0011		0011	0011	0011	0011
F	0101	0101	0101	0101		0101	0101	0101	0101
G	0000	0000	0010	0101		0000	0001	1100	0000

(b) Truth Table

$$G = \Sigma\, m(5, 10, 18, 22, 24, 42, 45, 47, 55, 56, 57)$$

(c) Equation for the Above Map

FIG. 6-6 Six-Variable Map

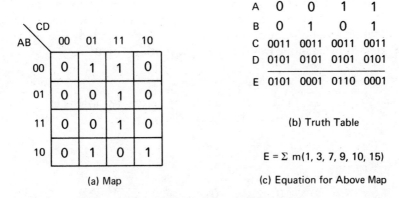

AB \ CD	00	01	11	10
00	0	1	1	0
01	0	0	1	0
11	0	0	1	0
10	0	1	0	1

(a) Map

A	0	0	1	1
B	0	1	0	1
C	0011	0011	0011	0011
D	0101	0101	0101	0101
E	0101	0001	0110	0001

(b) Truth Table

$E = \Sigma\, m(1, 3, 7, 9, 10, 15)$

(c) Equation for Above Map

FIG. 6-7 Map and Truth Table Shortcuts

Example 6-8: Chart and map the equation

$$Z = \sum m(9, 20, 21, 29, 30, 31)$$

SOLUTION:

Since the highest minterm is 31, this must be a five-variable problem. (Four variables go to m_{15}, five variables go to m_{31}, and six to m_{63}.) Charting,

A	0	0	0	0	1	1	1	1
B	0	0	1	1	0	0	1	1
C	0	1	0	1	0	1	0	1
D	0011	0011	0011	0011	0011	0011	0011	0011
E	0101	0101	0101	0101	0101	0101	0101	0101
Z	0000	0000	0100	0000	0000	1100	0000	0111

Mapping,

BC \ DE ($\overline{A}$)	00	01	11	10
00	0	0	0	0
01	0	0	0	0
11	0	0	0	0
10	0	1	0	0

(A)	00	01	11	10
00	0	0	0	0
01	1	1	0	0
11	0	1	1	1
10	0	0	0	0

Note that the designators have been omitted because the map order has been memorized.

Note that a four-bit all-0 group can be expressed by using a large 0 and a four-bit all-1 group by a large 1. Also, there is no need to copy the minterm designation numbers down, since their organization on the table can easily be determined by counting from left to right starting with 0.

Example 6-9: Chart and map:

$$R = \sum m(0, 1, 3, 5, 7, 10, 11, 21, 22, 23, 24, 26, 30, 32, 34, 35,$$
$$40, 41, 46, 47, 50, 51, 52, 60, 61)$$

SOLUTION:

Charting

A	0	0	0	0	0	0	0	0	1	1	1	1	1	1	1	1
B	0	0	0	0	1	1	1	1	0	0	0	0	1	1	1	1
C	0	0	1	1	0	0	1	1	0	0	1	1	0	0	1	1
D	0	1	0	1	0	1	0	1	0	1	0	1	0	1	0	1
E	0011	0011	0011	0011	0011	0011	0011	0011	0011	0011	0011	0011	0011	0011	0011	0011
F	0101	0101	0101	0101	0101	0101	0101	0101	0101	0101	0101	0101	0101	0101	0101	0101
G	1101	0101	0011	0000	0000	0111	1010	0010	1011	0000	1100	0011	0011	1000	0000	1100

Mapping,

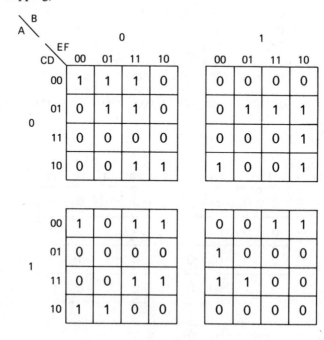

Example 6-10: Chart and map:

$$L = \bar{W}\bar{X}YZ + W\bar{X}Y\bar{Z} + W\bar{X}\bar{Y}\bar{Z} + \bar{W}XYZ$$

SOLUTION:

Converting to minterms:

$$L = \sum m(3, 10, 8, 7)$$

	CD 00	01	11	10
AB				
00	0	0	1	0
01	0	0	1	0
11	0	0	0	0
10	1	0	0	1

W   0   0   1   1

X   0   1   0   1

Y   0011 0011 0011 0011
Z   0101 0101 0101 0101
_____

L   0001 0001 1010 0000

6-4 SOLVING DIGITAL PROBLEMS USING MAPS

All of this mapping is no good unless it has some use. This technique can be used very effectively to reduce a Boolean expression to the minimum number of terms required for that expression. This result can then be implemented in **AND/OR**/invert logic. Therefore, the designer usually follows the following steps in solving a logic problem:

A. Expand the expression to minterms.
B. Map the minterms.
C. Reduce the map to the simplest product-of-sums (POS) form.
D. Read this product of sums from the map.
E. Reduce the map to the simplest sum-of-products (SOP) form.
F. Read this sum of products from the map.
G. Implement the simpler of D and F in **AND/OR**/invert logic.
H. Modify the result to obtain a **NAND** or **NOR** realization.

Note that both the SOP and POS expressions must be derived to see which is simpler.

6-5 SUM-OF-PRODUCTS MAP REDUCTION

Karnaugh maps (named after their inventor) are effective for reducing an expression to a sum-of-products form. In this section we shall discuss the theory and use of maps for this purpose.

Theory of Map Reduction

Most people don't like to operate in the dark. For this reason, let us discuss the scientific and mathematical bases for map reduction before applying it to the solution of problems.

Observe the map shown in Fig. 6-8. What it really says is

$$F = \sum m(0, 4, 10, 11, 14, 15)$$
$$= \bar{A}\bar{B}\bar{C}\bar{D} + \bar{A}B\bar{C}\bar{D} + A\bar{B}C\bar{D} + A\bar{B}CD + ABC\bar{D} + ABCD$$

CD \ AB	00	01	11	10
00	1_0	0_1	0_3	0_2
01	1_4	0_5	0_7	0_6
11	0_{12}	0_{13}	1_{15}	1_{14}
10	0_8	0_9	1_{11}	1_{10}

FIG. 6-8 Reducing to Sum-of-Products Form

Note that minterms 0 and 4 can be combined and reduced as follows:

$$\bar{A}\bar{B}\bar{C}\bar{D} + \bar{A}B\bar{C}\bar{D} = \bar{A}\bar{C}\bar{D}(\bar{B} + B) = \bar{A}\bar{C}\bar{D}(1) = \bar{A}\bar{C}\bar{D}$$

The reason this reduction was possible was that the two terms were identical except for one letter, B. But this is true for every horizontal or vertical adjacent pair on the map: minterms 4 and 5, minterms 13 and 5, and minterms 9 and 11, for example. In each case, the terms differ by only one literal. This is also true of combinations formed by wrapping the map around horizontally or vertically, for minterms 2 and 10 differ by only one element, as do minterms 8 and 0, 14 and 12, and 2 and 0.

Take a closer look at the map. See the binary numbers along the top? They differ by only one place when moving from left to right: 00, 01, 11, 10. The right number, 10, differs in only one place from the left number, 00, so they are also adjacent. Does this counting sequence look familiar? It is nothing more than a Gray code. Remember that the definition of a Gray code requires that each successive number differ only in one place. The map is therefore a device with a Gray code numbering system on both the vertical columns and horizontal rows. Although this text uses the maps shown, any map with a Gray code on both the horizontal and vertical will allow reduction.

Reading the Map

In the previous paragraphs, it was pointed out that minterms 0 and 4 were adjacent (called a *two-square*) and could be reduced. This can be done directly off the map by observing the characteristics of the 0 and 4 squares. For both squares,

$\bar{A}$ is an essential term, for the A part of the term is 0 in both terms. Similarly, both require $\bar{C}$ and $\bar{D}$. However, the B term can be either a 0 or a 1. Therefore, collecting only those essential terms, the reduced result is $\bar{A}\bar{C}\bar{D}$, which is identical to the Boolean algebra reduction method. Look at terms 10, 11, 14, and 15 on the map. These terms form a perfect square called a *four-square* and can be reduced using Boolean algebra as follows:

$$F = \sum m(10, 11, 14, 15)$$
$$= A\bar{B}C\bar{D} + A\bar{B}CD + ABC\bar{D} + ABCD$$
$$= A\bar{B}C(\bar{D} + D) + ABC(\bar{D} + D)$$
$$= A\bar{B}C(1) + ABC(1)$$
$$= A\bar{B}C + ABC$$
$$= AC(\bar{B} + B)$$
$$= AC(1)$$
$$F = AC$$

These four terms can also be reduced directly off the map by observing which terms are essential to all four-squares:

A. A is a 1 in all cases.
B. B can be a 1 or a 0.
C. C is a 1 in all cases.
D. D can be a 1 or a 0.

Therefore, collecting the essential terms, the reduced term is AC, and we can state that AC "covers" minterms 10, 11, 14, and 15.

From the foregoing paragraphs, the reduction of the entire expression mapped in Fig. 6-8 is

$$F = \sum m(0, 4, 10, 11, 14, 15)$$
$$F = \bar{A}\bar{C}\bar{D} + AC$$

This could next be implemented in **AND/OR**/invert logic and then converted to **NAND** or **NOR** logic, as discussed in Chapter 4.

Selection of Squares

Using these same principles, terms on a two-, three-, or four-variable map can be reduced if they satisfy the following conditions:

A. They must form 2^n-squares (two-squares, four-squares, eight-squares, sixteen-squares, . . .).
B. They must form a geometric square or rectangle. They can be formed by wrapping around but cannot be formed using diagonal configurations.

Some examples of two-squares include minterms 0 and 1, 1 and 3, 0 and 2, and 4 and 6. Some examples of four-squares include 0, 1, 2, 3; 0, 1, 4, 5; 0, 2, 4, 6; 1, 5,

9, 13; 1, 3, 9, 11; and 0, 2, 8, 10. Some examples of eight-squares include 0, 1, 2, 3, 4, 5, 6, 7; 0, 1, 2, 3, 8, 9, 10, 11; 4, 5, 6, 7, 12, 13, 14, 15; and 0, 2, 4, 6, 8, 10, 12, 14.

Reduction of Boolean Expressions

Let us now state some rules for map reduction:

A. Select first that term that has the fewest number of adjacencies.
B. Expand this expression to include the largest number of squares possible in a 2^n configuration. Squares may be covered several times if necessary for this expansion.
C. Continue to select uncovered terms that have the fewest number of adjacencies.
D. All terms must be covered in the final expression.
E. Finally, read the map. A two-square will eliminate one variable, a four-square two variables, an eight-square three variables, and a 2^n-square, n variables.

Example 6-11: Reduce the expression $F = \sum m(0, 1, 2, 3, 6, 7, 13, 15)$ by mapping and implement in **NAND** logic.

SOLUTION:

Plot on a four-variable map (Fig. 6-9). Minterm 13 has the fewest number of adjacencies; it can only be combined with m_{15}. Note that m_3 has three adjacencies: m_1, m_2, and m_7. Therefore, expanding m_{13}, all we get is a two-square with m_{15}. Mark as shown on the map.

Minterm 0, if it is chosen and expanded, can only go with m_1, m_2, and m_3 to form a four-square. Thus, we are forced to include this combination in the final expression. Similarly, m_6 can only go with m_2, m_3, and m_7 to form a four-square. Note that m_2 and m_3 are covered twice. Although no effort should be made to cover a term twice, it may be done in order to expand the expression to the largest number of squares.

The next process is to read the map. Reading the top four-square (m_0, m_1, m_2, and m_3) yields $\bar{A}\bar{B}$, for C and D can both be 1's and 0's. Reading m_2, m_3, m_6, and m_7 produces $\bar{A}C$, and reading m_{13} and m_{15} gives ABD. The final expression is therefore

$$F = \bar{A}\bar{B} + \bar{A}C + ABD$$

It then can be implemented in **AND/OR**/invert logic, and then **NAND** logic according to the procedures shown in Chapter 4. Figure 6-9 illustrates this implementation.

Example 6-12: Reduce and implement in **NOR** logic: $F = \sum m(2, 3, 5, 7, 9, 11, 12, 13, 14, 15)$.

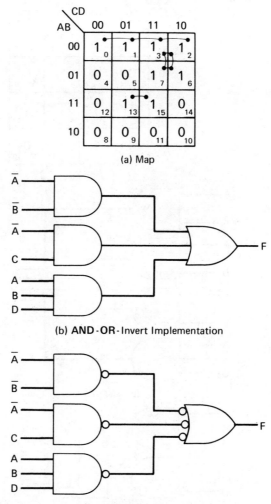

(a) Map

(b) **AND-OR-**Invert Implementation

FIG. 6-9 Example 6-11   (c) **NAND** Implementation

SOLUTION:

Mapping the function yields Fig. 6-10. In selecting the necessary combinations, note the following:

A. m_{15} can go with three different four-squares, so do not select it yet.

B. m_9, expanded, can go with only one four-square, so select it (m_9, m_{11}, m_{13}, m_{15}).

C. m_{12}, expanded, can go with only one four-square, so select it (m_{12}, m_{13}, m_{14}, m_{15}).

D. m_2 expands to only one two-square, so choose it (m_2, m_3).

E. This leaves only m_5 and m_7 uncovered. m_7 can go with two different

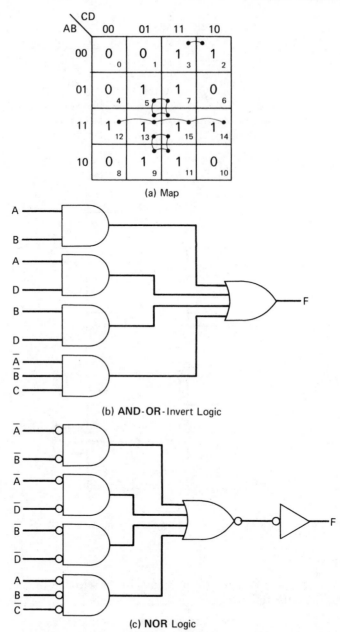

(a) Map

(b) **AND**-**OR**-Invert Logic

(c) **NOR** Logic

FIG. 6-10 Example 6-12

four-squares, but m_5 can only go with one. Therefore, choose m_5, expanded, making four-square m_5, m_7, m_{13}, m_{15}.

F. All 1's are now covered. Note that it was not necessary to choose four-square m_3, m_7, m_{11}, m_{15}.

Reading the map produces the following terms:

A. Two-square m_2, m_3 reduces to $\bar{A}\bar{B}C$.

B. Four-square m_5, m_7, m_{13}, m_{15} reduces to BD.

C. Four-square $m_{12}, m_{13}, m_{14}, m_{15}$ reduces to AB.

D. Four-square $m_9, m_{11}, m_{13}, m_{15}$ reduces to AD.

Therefore, $F = \bar{A}\bar{B}C + BD + AB + AD$. Figure 6-10 shows the **NOR** implementation of the function.

Example 6-13: Reduce $F = \sum m(0, 2, 3, 4, 5, 7)$ using mapping.

SOLUTION:

Figure 6-11 illustrates the map of the function. However, we immediately run into a problem, for each minterm has two adjacencies. What this means is that there may be more than one final minimum expression. Therefore, we shall arbitrarily choose two-square m_0, m_2. Using this scheme, m_3 combines with m_7 and m_4 with m_5. Therefore, the final expression will be three two-squares, as shown in Fig. 6-11(a). But what if we had chosen m_0, m_4? Then m_5 would combine with m_7 and m_3 with m_2, giving three two-squares for the answer, as shown in Fig. 6-11(b). Both expressions have the same number of gates and are equally acceptable.

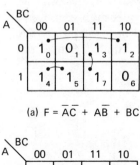

(a) $F = \bar{A}\bar{C} + A\bar{B} + BC$

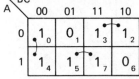

FIG. 6-11 Example 6-13   (b) $F = \bar{B}\bar{C} + AC + \bar{A}B$

Counting the Cost

In the previous paragraph I made the rash statement that there were two solutions to a problem that were equally acceptable. "How can he be so sure of himself?" you ask. You are about to find out.

When logic was executed using discrete components, the criterion for the optimum solution was how many components, usually diodes, were included in the finished circuit. With the advent of integrated circuits a similar criterion is used: how many inputs are required? Although not perfect, it is a reasonable assumption considering that the cost of each input is about the same. Table 6-2 lists four integrated circuits of equal cost showing the number of gates each contains and the number of input leads (total gates times number of inputs per gate). As can be seen, each IC has about the same number of inputs per package: therefore, each input has about equal cost.

TABLE 6-2. Integrated Circuit Gates

Type IC	Type Gate	Number of Gates	Total Inputs
7400	Two-input **NAND**	4	8
7410	Three-input **NAND**	3	9
7420	Four-input **NAND**	2	8
7430	Eight-input **NAND**	1	8

So much for a criterion. But how does one determine how many inputs a function will have? First let us examine a typical sum-of-products equation:

$$Y = A\bar{B}C + \bar{A}\bar{B}\bar{C} + \bar{A}B + \bar{A}C$$

As shown in Fig. 6-12, each **AND** term requires one input for each letter shown: three for $A\bar{B}C$, three for $\bar{A}\bar{B}\bar{C}$, two for $\bar{A}B$, and two for $\bar{A}C$, making a total of 10 inputs. But the outputs of the **AND** gates must each feed an **OR** gate. Since there are four **AND** gates, the **OR** gate must have 4 inputs. Adding this to the previous 10 inputs means that 14 total inputs are required. Similar procedures can be used for the product-of-sums expression.

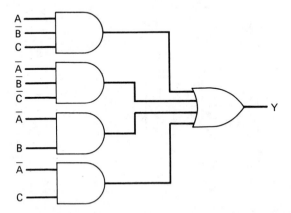

FIG. 6-12 Counting Gate Inputs

Example 6-14: How many inputs are required for the following equations?

(a) $W = A\bar{B}D + AC\bar{D} + EF$

(b) $X = LM + \bar{N}PQ + L\bar{M}P\bar{Q}$

(c) $\quad Y = S\bar{T}\bar{U}V + ST\bar{O}V + U\bar{V} + SUV + T\bar{U}\bar{V}$

(d) $\quad Z = (A + \bar{B} + C)(\bar{A} + D)(B + \bar{D})$

SOLUTION:

(a) Write the equation $\qquad\qquad W = A\bar{B}D + AC\bar{D} + EF$

Count **AND** inputs	3	+	3	+	2	=	8
Count **AND** gates feeding **OR** gates	1	+	1	+	1	=	3

Total inputs $\qquad\qquad\qquad\qquad\qquad\qquad\qquad$ 11

(b) Write the equation $\qquad\qquad X = LM + \bar{N}PQ + L\bar{M}P\bar{Q}$

Count **AND** inputs	2	+	3	+	4	=	9
Count **OR** inputs	1	+	1	+	1	=	3

Total inputs $\qquad\qquad\qquad\qquad\qquad\qquad\qquad$ 12

(c) Write equation $\quad Y = S\bar{T}\bar{U}V + ST\bar{O}V + U\bar{V} + SUV + T\bar{U}\bar{V}$

AND inputs	4	+	4	+	2	+	3	+	3	= 16
OR inputs	1	+	1	+	1	+	1	+	1	= 5

Total inputs $\qquad\qquad\qquad\qquad\qquad\qquad\qquad$ 21

(d) Write equation $\qquad\qquad Z = (A + \bar{B} + C)(\bar{A} + D)(B + \bar{D})$

OR inputs	3	+	2	+	2	=	7
AND inputs	1	+	1	+	1	=	3

Total inputs $\qquad\qquad\qquad\qquad\qquad\qquad\qquad$ 10

Five-and Six-Variable Reduction

Five- and six-variable maps have more than one block of 16 squares and may contain two-squares, four-squares, or other combinations involving these blocks. Squares are considered adjacent in two blocks if, when superimposing one block on top of another block that is above or beside the first block, the squares coincide with one another. For example, in Fig. 6-13, minterms 1 and 17, 47 and 63, and 28 and 60 are each two-squares. Note that each differs by exactly 2^n; this is also required in order to be adjacent. Diagonal elements such as m_5 and m_{53} are not adjacent. The following are examples of four-squares in a six-variable field:

A. m_5, m_7, m_{21}, m_{23}

B. m_4, m_6, m_{36}, m_{38}

C. $m_{11}, m_{27}, m_{43}, m_{59}$

This procedure can be extended to any 2^n-sized squares.

Reduction is accomplished by observing which variable or variables drop out. For example, the four-square m_4, m_6, m_{36}, m_{38} straddles $A = 0$ and $A = 1$, so the A variable drops out, leaving $\bar{B}\bar{C}D\bar{F}$. Similarly, four-square $m_{11}, m_{27}, m_{43}, m_{59}$ drops both the A and B variables because it includes $A = 0$, $A = 1$, $B = 0$, and $B = 1$, leaving $C\bar{D}EF$. Reading the other reduced terms from the map,

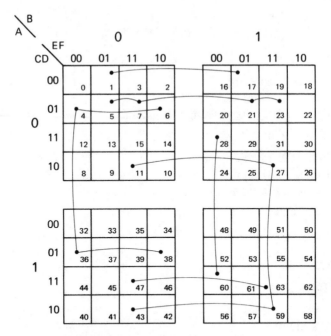

FIG. 6-13 Six-Variable Adjacencies

A. $m_1, m_{17} = \bar{A}\bar{C}\bar{D}\bar{E}F$

B. $m_{47}, m_{63} = ACDEF$

C. $m_{28}, m_{60} = BC D\bar{E}\bar{F}$

D. $m_5, m_7, m_{21}, m_{23} = \bar{A}\bar{C}DF$

Example 6-15: Reduce the following equation using mapping:

$$U = \sum m(2, 3, 6, 7, 9, 10, 12, 13, 14, 16, 18, 22, 26, 28, 29, 30)$$

SOLUTION:

Figure 6-14 is a map of the minterms. m_9 can only be paired with m_{13}; m_{16} can only go with m_{18}; m_{28} can go with m_{12}, m_{14}, m_{30}, or it can go with m_{12}, m_{13}, m_{29}, so do not select it yet. m_{29} can only go with m_{28}, m_{12}, m_{13}, so select this four-square. m_7 can only go with four-square m_2, m_3, m_6. The remaining terms will be covered by eight-square $m_2, m_6, m_{10}, m_{14}, m_{18}, m_{22}, m_{26}, m_{30}$. The final reduced equation from the map is

$$U = \bar{A}B\bar{D}E + A\bar{B}\bar{C}\bar{E} + \bar{A}\bar{B}D + BC\bar{D} + D\bar{E}$$

Example 6-16: Reduce the following equation using mapping:

$$Y = \sum m(0, 2, 4, 8, 10, 13, 15, 16, 18, 20, 23, 24, 26, 32, 34,$$
$$40, 42, 45, 47, 48, 50, 56, 57, 58, 60, 61)$$

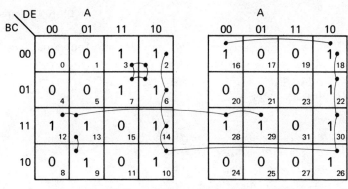

$$U = \overline{AB}\overline{D}\overline{E} + \overline{A}\overline{B}D + D\overline{E} + BC\overline{D} + A\overline{B}\overline{C}\overline{E}$$

FIG. 6-14 Example 6-15

SOLUTION:

Figure 6-15 is a map of the minterms. All the terms have at least one adjacency except for m_{23}, so select m_{23}. Next, m_4 can only go with four-square

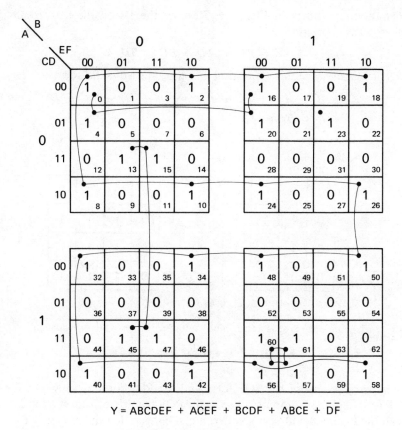

$$Y = \overline{AB}\overline{C}DEF + \overline{A}\overline{C}\overline{E}\overline{F} + \overline{B}CD\overline{F} + ABC\overline{E} + \overline{D}\overline{F}$$

FIG. 6-15 Example 6-16

149

m_0, m_4, m_{16}, m_{20}. Similarly, m_{13} can only go with four-square $m_{13}, m_{15}, m_{45}, m_{47}$. m_{60} can only go with four-square $m_{56}, m_{57}, m_{60}, m_{61}$. m_8 goes with sixteen-square $m_0, m_2, m_8, m_{10}, m_{16}, m_{18}, m_{24}, m_{26}, m_{32}, m_{34}, m_{40}, m_{42}, m_{48}, m_{50}, m_{56}, m_{58}$. Reading the map,

A. m_{23} $\quad\quad\quad\quad\quad = \bar{A}B\bar{C}DEF$
B. m_0, m_4, m_{16}, m_{20} $\quad = \bar{A}\bar{C}E\bar{F}$
C. $m_{13}, m_{15}, m_{45}, m_{47}$ $\quad = \bar{B}CDF$
D. $m_{56}, m_{57}, m_{60}, m_{61}$ $\quad = ABC\bar{E}$
E. $m_0, m_2, \ldots, m_{56}, m_{58} = \bar{D}\bar{F}$

The final result is

$$Y = \bar{A}B\bar{C}DEF + \bar{A}\bar{C}E\bar{F} + \bar{B}CDF + ABC\bar{E} + \bar{D}\bar{F}$$

6-6 PRODUCT-OF-SUMS REDUCTION

Consider the map shown in Fig. 6-16. Reading the 1's off the map results in the sum-of-products expression

$$Z = AC + AD + BC + BD$$

However, let us massage this expression into a product-of-sums form:

$$Z = AC + AD + BC + BD = A(C + D) + B(C + D)$$
$$= (C + D)(A + B)$$

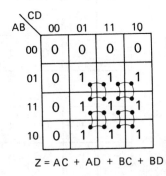 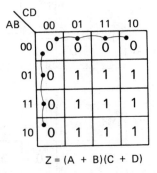

Z = AC + AD + BC + BD   Z = (A + B)(C + D)

(a) Sum – of – Products Reduction   (b) Product – of – Sums Reduction

FIG. 6-16 POS and SOP Map Reduction

Now, compare the cost of the SOP form and the POS form as shown in Fig. 6-17. The SOP form requires 12 inputs, whereas the POS requires 6, a substantial difference. Although the POS is not always simpler, it can be seen that the only way to perform an effective cost analysis is to compare the POS and SOP forms.

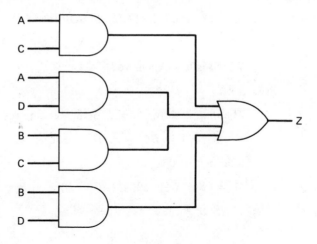

(a) Sum-of-Products Form

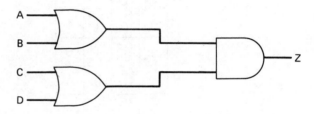

FIG. 6-17 SOP and POS Logic   (b) Product-of-Sums Form

Take another look at the two maps (Fig. 6-16). The POS form required reading the 1's off the map, necessitating forming the four four-squares shown. But look how neatly the 0's form one vertical column and one horizontal row. Such a tasty morsel of four-squares! If only they were not 0's, but 1's. But take heart; I have good news for you—you can read the 0's to form the POS form rather than the SOP form.

To read the POS form,

A. Combine the 0's into 2^n-squares, just as you would the 1's.
B. Read the 0's from the map and record the variables.
C. To form the sums, change the **AND**s to **OR**s and complement the individual variables.
D. Form the product (**AND**) of these sums.

Referring to Fig. 6-16 again, read the 0's from the map, just as you would the 1's, obtaining $\bar{A}\bar{B}$ and $\bar{C}\bar{D}$. Next, perform step C, changing $\bar{A}\bar{B}$ to $A + B$ and $\bar{C}\bar{D}$ to $C + D$. $A + B$ and $C + D$ are the sums, and all that needs to be done is to form the product (**AND**) of these sums:

$$Y = (A + B)(C + D)$$

Example 6-17: Find the POS and SOP forms of

$$Y = \sum m(0, 1, 3, 6, 7, 8, 9, 13, 15)$$

Which is less expensive?

SOLUTION:

The map is shown in Fig. 6-18. Reading the 1's,

$$Y = \bar{B}\bar{C} + \bar{A}\bar{B}D + \bar{A}BC + ABD \quad \text{(15 inputs)}$$

Reading the 0's,

$$Y = (B + \bar{C} + D)(\bar{A} + \bar{B} + D)(A + \bar{B} + C)(\bar{A} + B + \bar{C}) \quad \text{(16 inputs)}$$

The SOP form is less expensive.

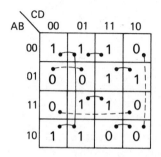

$Y = \bar{B}\bar{C} + \bar{A}\bar{B}D + \bar{A}BC + ABD$
$= (B + \bar{C} + D)(\bar{A} + \bar{B} + D)(A + \bar{B} + C)(\bar{A} + B + \bar{C})$ **FIG. 6-18** Example 6-17

Example 6-18: Find the minimum expression for the equation

$$Y = \sum m(2, 3, 4, 5, 6, 7, 12, 13, 14, 15, 18, 19, 20, 21, 22, 23, 28, 29, 30, 31)$$

SOLUTION:

Plotting on a five-variable map (Fig. 6-19) and reading the 1's for the SOP form,

$$Y = C + \bar{B}D$$

Reading the 0's for the POS form,

$$Y = (\bar{B} + C)(C + D)$$

The SOP form requires four inputs (note that there is no need for an **AND** gate for the C term); the POS form requires six. Therefore, the minimum expression is

$$Y = C + \bar{B}D$$

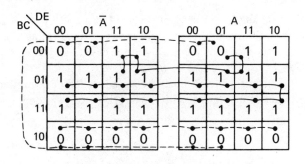

$$Y = C + \bar{B}D$$
$$= (\bar{B} + C)(C + D)$$

FIG. 6-19 Example 6-18

Maxterms

Let's be logical about logic. If there is an animal called a minterm, there surely must be one called a maxterm. There is. A maxterm is a sum of all the variables within the logic system. If A, B, C, D, and E are all the variables within a system, then the following would represent maxterms:

A. $A + B + \bar{C} + D + E$
B. $\bar{A} + \bar{B} + C + \bar{D} + E$
C. $A + B + C + D + E$
D. $A + B + \bar{C} + D + \bar{E}$
E. $\bar{A} + \bar{B} + \bar{C} + \bar{D} + \bar{E}$

Just as any Boolean expression can be transformed into a sum of minterms, any Boolean expression can be written as a product of maxterms.

Maxterms are plotted as 0's on a map for the purpose of reduction. However, the numbering of maxterms is slightly different from the numbering of minterms. Referring to Fig. 6-20, read the square labeled "14" off the map as a sum: $\bar{A} + \bar{B} +$

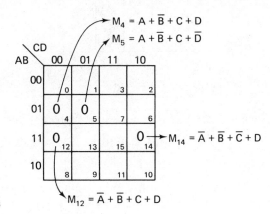

FIG. 6-20 Maxterms

$\bar{C} + D$. As can be seen, maxterm 14 is $\bar{A} + \bar{B} + \bar{C} + D$ with the **NOT** literals called 1's and the uncomplemented literals 0's in forming the binary maxterm designators. Maxterm 4 (abbreviated M_4 with a capital M) is $A + \bar{B} + C + D$, M_5 is $A + \bar{B} + C + \bar{D}$, and M_{12} is $\bar{A} + \bar{B} + C + D$.

Just as a problem can be stated as a sum of minterms, it can also be stated as a product of maxterms, with the Greek letter $\prod$, upper case pi, representing a product:

$$Y = \prod M(0, 1, 3, 4)$$
$$= (A + B + C)(A + B + \bar{C})(A + \bar{B} + \bar{C})(\bar{A} + B + C)$$

These 0's can then be plotted on a map, the rest of the squares set to 1, and then reduced to SOP or POS forms.

Example 6-19: Find the minimum expression of

$$Y = \prod M(0, 1, 3, 5, 6, 7, 10, 14, 15)$$

SOLUTION:

Plotting these maxterms as 0's on the map and the remaining squares as 1's results in Fig. 6-21. Reading the POS form,

$$Y = (A + B + C)(\bar{A} + \bar{C} + D)(A + \bar{D})(\bar{B} + \bar{C})$$

resulting in 14 inputs. Reading the SOP form,

$$Y = \bar{A}\bar{B}C\bar{D} + B\bar{C}\bar{D} + A\bar{B}D + A\bar{C}$$

resulting in 16 inputs. Therefore, the POS form is less expensive.

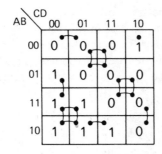

$Y = (A + B + C) (\bar{A} + \bar{C} + D) (A + \bar{D}) (\bar{B} + \bar{C})$
$Y = \bar{A}\bar{B}C\bar{D} + B\bar{C}\bar{D} + A\bar{B}D + A\bar{C}$ **FIG. 6-21** Example 6-19

Example 6-20: Find the minimum expression for

$$Y = \prod M(0, 1, 9, 10, 11, 13, 14, 15, 16, 17, 22, 23, 26, 27)$$

SOLUTION:

When dealing with complex maps, it helps to draw separate maps for reading the SOP and POS forms (Fig. 6-22).

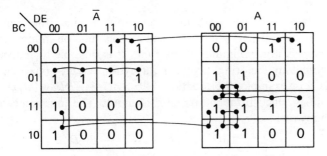

$$Y = \overline{B}\overline{C}D + \overline{A}\overline{B}C + ABC + B\overline{D}\overline{E} + AC\overline{D} + AB\overline{D}$$

(a) SOP Map

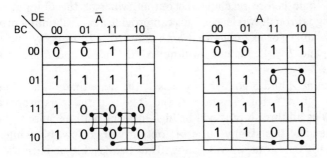

$$Y = (B + C + D)(\overline{B} + C + \overline{D})(\overline{A} + B + \overline{C} + \overline{D})(A + \overline{B} + \overline{E})(A + \overline{B} + \overline{D})$$

(b) POS Map

FIG. 6-22 Example 6-20

Reading the SOP map,

$$Y = \bar{B}\bar{C}D + \bar{A}\bar{B}C + ABC + B\bar{D}\bar{E} + AC\bar{D} + AB\bar{D}$$

requiring 24 inputs. Reading the POS map,

$$Y = (B + C + D)(\bar{B} + C + \bar{D})(\bar{A} + B + \bar{C} + \bar{D})$$
$$\times (A + \bar{B} + \bar{E})(A + \bar{B} + \bar{D})$$

requiring 21 inputs. Therefore, the POS form is the minimum expression.

6-7 HYBRID FUNCTIONS

Sum-of-products and product-of-sums reduction produces an input gate feeding and an output gate. Therefore, it is called *two-level logic*, since each input signal has to pass through two gates before reaching the output. This has the advantage of providing uniform time delay between input signals and the output; if each gate has a 10-ns delay, all input signals are delayed 20 ns between input and output.

However, two-level logic does not necessarily produce the simplest circuit. How's that for an eye opener! Here we spend all this time on minimization only to find out minimum is not minimum. Many times it is possible to reduce a Boolean SOP or POS circuit by factoring, producing a hybrid circuit—one that has greater than two levels. For example, the equation $Y = ABC + ABD + ACD + BCD$ is a minimum sum-of-products form and requires 16 inputs. The equation can be reduced by factoring, however;

$$Y = ABC + ABD + ACD + BCD$$
$$= AB(C + D) + CD(A + B)$$

Implementing this result (Fig. 6-23) shows we have reduced the number of inputs from 16 to 12. Note, however, that the C input to the **OR** gate must go through three levels of logic before reaching the output, whereas the C input to the **AND** gate must only go through two levels. This can provide a critical timing problem called a *logic race*. Assume, for example, that each gate has a 10-ns delay and that $A = 0$, $B = 0$, $C = 1$, and $D = 1$. Gate A_2 will not **AND** since A and B are 0; gate A_4 will not **AND** since $A + B = 0$.

Next, assume A and B go high at precisely the same time C and D go low. Gate A_1 will provide a 1 to A_2 for 10 ns after C and D go to 0 due to its propagation delay, and for that 10 ns all three inputs will be high, causing a 10-ns pulse to be outputted by A_5. At the end of this skinny pulse, A_1 output will go low, blocking A_2; since C and D are already 0, A_5 output will go low. Had two-level logic been used, this logic race and its resulting pulse would not have occurred.

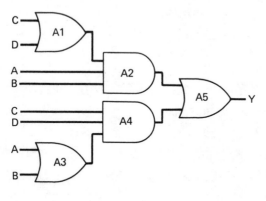

$$Y = AB(C + D) + CD(A + B)$$ **FIG. 6-23** Hybrid Logic

Decoders

A decoder is a circuit that will energize a particular line or lines, depending on the binary code at the input. One such device decodes binary data and energizes a particular line at the output specified by the binary code. Figure 6-24(a) illustrates a circuit that has 16 output lines. A binary 12 on the input would energize line 12, and

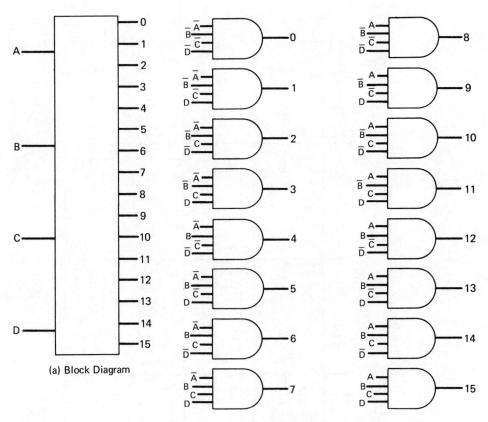

(a) Block Diagram

(b) Using Four Input Gates

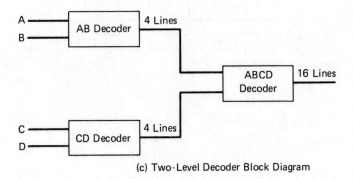

(c) Two-Level Decoder Block Diagram

FIG. 6-24 One of 16 Decoder

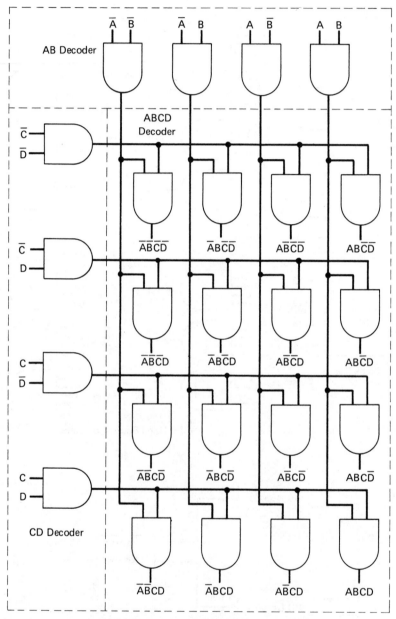

(d) Two Level Decoder Logic Diagram

FIG. 6-24 (continued)

a binary 7, line 7. One possible manner in which this might be accomplished is to use 16 four-input gates, each decoding one binary input [Fig. 6-24(b)]. This would require 16×4 or 64 inputs. However, by converting this to a two-level function, we can save some money. To do this, we shall build two decoders: an AB decoder (a 1 of 4 decoder) and a CD decoder (a second 1 of 4 circuit). These two circuits can then feed second-level output gates [Fig. 6-24(c)]. The completed logic diagram is shown in Fig. 6-24(d). Note that the total input count is

AB decoder	$4 \times 2 =$	8
CD decoder	$4 \times 2 =$	8
Output gates	$16 \times 2 =$	32
Total		48

This compares with 64 if we had used the four input gates.

This concept can be extended to any number of inputs. The general approach is as follows:

A. List all the input variables.
B. Break the list roughly in half, and continue breaking the resultant groups until groups of two or three variables are obtained.
C. Design decoders for each group.

Example 6-21: Design a 1 of 512 decoder using multilevel logic.

SOLUTION:

The output variables would be $ABCDEFGHI$. Grouping these,

$$(ABCDE)(FGHI)$$

and again,

$$[(ABC)(DE)][(FG)(HI)]$$

The block and logic diagrams for the circuit are shown in Fig. 6-25. Each decoder is a combinational logic circuit consisting of two or three input gates. Because the logic diagram is quite extensive, only the $FGHI$ decoder is shown. If a single-level decoder had been used, the gate count would have been

$$9 \times 512 = 4608$$

For this two-level circuit, the count is

ABC decoder	$8 \times 3 =$	24
DE decoder	$4 \times 2 =$	8
FG decoder	$4 \times 2 =$	8
HI decoder	$4 \times 2 =$	8
$ABCDE$ decoder	$32 \times 2 =$	64
$FGHI$ decoder	$16 \times 2 =$	32
Output gates	$512 \times 2 =$	1024
Total		1168

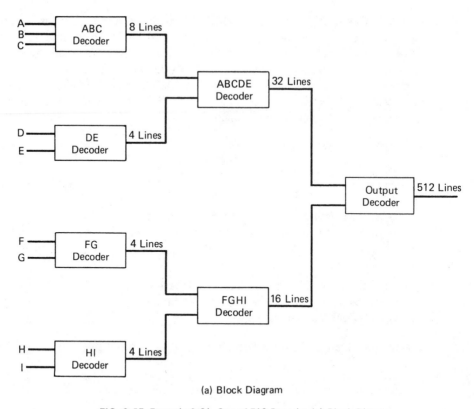

(a) Block Diagram

FIG. 6-25 Example 6-21, One of 512 Decoder (a) Block Diagram

Parity Detectors

Another very useful multilevel function is the parity detector-generator. Figure 6-26 illustrates the circuit, which consists of cascaded exclusive **OR** gates. Remember that an exclusive **OR** gate will provide an output of 1 if either input (but not both) is a 1. Therefore, we could state that if the number of 1's on its input is even, it will output a 0; if odd, it will output a 1. When the gates are connected as shown, they can detect or generate parity, for when the number of 1's at the input is even, the output is a 0. When the number of 1's at the input is odd, the output is a 1.

The circuit shown can be used either to generate parity at the transmitter or to detect parity at the receiver. If used at the transmitter, bits *A*, *B*, *C*, and *D* are data bits and bit *E* the parity bit. If odd parity is desired, bit *E* can be inverted.

When used at the receiver, the circuit detects parity over four bits, *A*, *B*, *C*, and

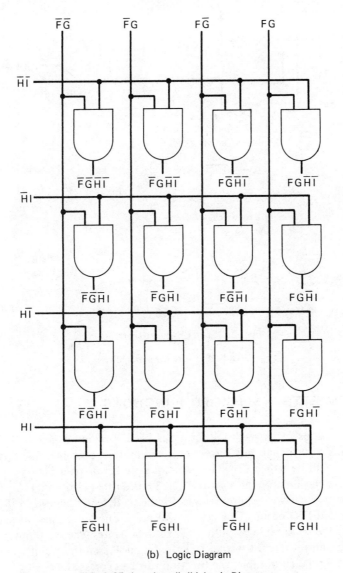

(b) Logic Diagram

FIG. 6-25 (continued) (b) Logic Diagram

D, and provides an error signal, E, if the parity is not even. If odd parity is used, E is merely inverted, and the error signal occurs when parity is even.

Although a four-bit parity detector-generator is shown, any word length can be handled by connecting additional **XOR** gates in cascade. It should be noted, though, that this increases propagation delay and logic race problems.

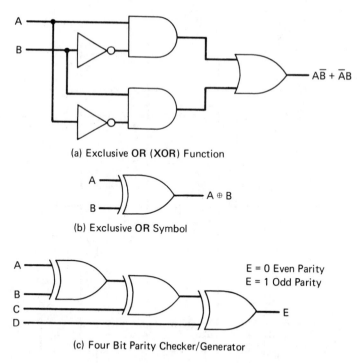

(a) Exclusive OR (XOR) Function

(b) Exclusive OR Symbol

(c) Four Bit Parity Checker/Generator

FIG. 6-26 Parity Generator/Checker

6-8 INCOMPLETELY SPECIFIED FUNCTIONS

In all the previous maps, each minterm has been specified as a 1 or a 0. However, many designs require that only certain of the minterms (or maxterms) be defined, and any others could be either 1's or 0's. An example of this is a binary-coded decimal system where the binary states 1010, 1011, 1100, 1101, 1110, and 1111 never occur.

Let us assume we require a logic circuit to detect whenever a *BCD* number higher than 5 (0101) occurs. The circuit will have four input lines, as shown in Fig. 6-27. Define the truth table by putting 0's in those states where a 0 output is desired, 1's on those states where a 1 output is desired, and X's on those states that will never occur. Reading the truth table, the function can be expressed as

$$Y = \sum m(6, 7, 8, 9) + d(10, 11, 12, 13, 14, 15)$$

where the d terms represent the unspecified states: those that will never occur. These terms are also called the *don't cares*.

Next, transfer this information to the map. Then read the SOP form off the map using the X's as 1's whenever needed to make a larger 2^n-square as shown. This results in

$$Y = A + BC$$

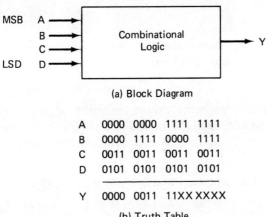

(a) Block Diagram

A	0000	0000	1111	1111
B	0000	1111	0000	1111
C	0011	0011	0011	0011
D	0101	0101	0101	0101
Y	0000	0011	11XX	XXXX

(b) Truth Table

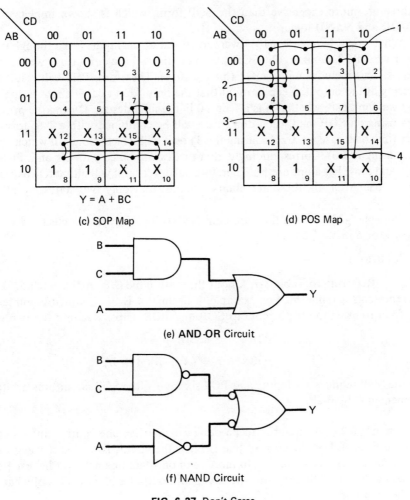

$Y = A + BC$

(c) SOP Map

(d) POS Map

(e) AND-OR Circuit

(f) NAND Circuit

FIG. 6-27 Don't Cares

By using the don't cares as shown, the result was a less expensive circuit, for if these had been 0's, the result would have been

$$Y = \bar{A}BC + A\bar{B}\bar{C}$$

However, the POS form must also be evaluated. Maybe there is an even less expensive circuit. Therefore, combine the X's with the 0's on the POS map, whenever helpful, and read the result. In this particular problem, there are three equally expensive POS forms:

Using four-squares 1 and 2 $Y = (A + B)(A + C)$

Using four-squares 2 and 4 $Y = (B + \bar{C})(A + C)$

Using four-squares 1 and 3 $Y = (A + B)(\bar{B} + C)$

All three are more expensive than the SOP form, which is shown implemented in **AND/OR** and **NAND** logic.

What would happen if a 1011 were to appear at the inputs (even though we said it never would)? Examining the SOP map, note that all the don't cares were assumed to be 1's. This includes minterm 11 (binary 1011). Therefore, 1011 will result in a 1 appearing at the output. Keep in mind that the only "don't care" about a don't care is that we don't care whether it is a 1 or a 0 in the final design. During the process of design using an SOP map, each don't care will be combined with a 1 to form a 2^n-square (in which case we are declaring it a 1) or it will be left alone (in which case we are declaring it a 0). Correspondingly, don't cares combined with 0's on a POS map are 0's, and those left alone are 1's. Therefore, all don't cares will ultimately be resolved to 1's or 0's. After all, these are binary systems, and only two states are allowed.

Example 6-22: Design the minimum **NAND** logic circuit to detect a decimal 3, 4, 5, or 6 in 5211 code.

SOLUTION:

Referring to Table 3-1, fill out the truth table (Fig. 6-28). Each 5211 code represents a minterm. For example, a decimal 4 is 0111, which is minterm 7. Then transfer the truth table to SOP and POS maps. Reading the two maps,

SOP $Y = \bar{A}B + A\bar{B}$

POS $Y = (A + B)(\bar{A} + \bar{B})$

Since the SOP requires six inputs and POS six inputs, the SOP circuit was arbitrarily implemented (Fig. 6-28).

Example 6-23: A circuit is to be designed that has one control line and three data lines. When the control line is high, the circuit is to detect when one of the data lines has a 1 on it. No more than one data line will ever have a 1 on it. When the control line is low, the circuit will output a 0, regardless of what is on the data lines.

```
MSB   A   0000  0000  1111  1111
      B   0000  1111  0000  1111
      C   0011  0011  0011  0011
LSB   D   0101  0101  0101  0101
      ──────────────────────────────
      Y   00X0  X1X1  1X1X  0X00
```

(a) Truth Table

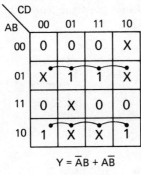

$$Y = \overline{A}B + A\overline{B}$$

(b) SOP Map

$$Y = (A + B)(\overline{A} + \overline{B})$$

(c) POS Map

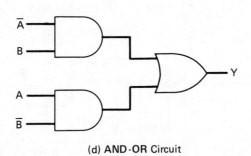

(d) AND-OR Circuit

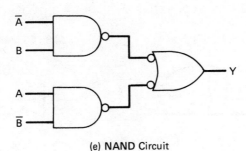

(e) NAND Circuit

FIG. 6-28 Example 6-22

SOLUTION:

A block diagram for the circuit is shown in Fig. 6-29(a), with C being the control line and D_1, D_2, and D_3 the data lines. Next, the truth table is shown, with X's (don't cares) entered anytime more than one data line is high, 1's anytime C is high and one of the three data lines is high, and 0's for all other minterms. I have to confess that when I first drew the truth table, I incorrectly

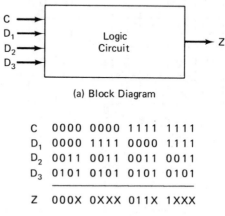

(a) Block Diagram

C	0000	0000	1111	1111
D_1	0000	1111	0000	1111
D_2	0011	0011	0011	0011
D_3	0101	0101	0101	0101

| Z | 000X | 0XXX | 011X | 1XXX |

(b) Truth Table

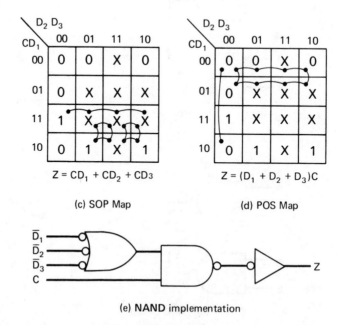

$Z = CD_1 + CD_2 + CD_3$

(c) SOP Map

$Z = (D_1 + D_2 + D_3)C$

(d) POS Map

(e) NAND implementation

FIG. 6-29 Example 6-23

put 0's on Z anytime C was 0. However, the more don't cares that are used, the simpler the final circuit. Therefore, I corrected the table to that shown.

Next, POS and SOP maps are constructed from the truth table. The two results are

$$Z = CD_1 + CD_2 + CD_3$$
$$= C(D_1 + D_2 + D_3)$$

Since the POS is simpler, it is implemented in **NAND** logic. The inputs to the **OR** gate have been negated, and by the time they pass through the inverter on the input to the gate, they will appear noninverted, as the equation specifies.

6-9 MULTIPLE-OUTPUT MINIMIZATION

Up to this time, we have only considered logic systems that have multiple inputs and one output. However, there are many design problems requiring several outputs, as shown in Fig. 6-30. One example is of a dual output function where input lines A, B, and C are to provide two output lines as defined by W and X in the truth table (Fig. 6-31). The maps for W and X are shown in Fig. 6-31(c) and (d). The results from

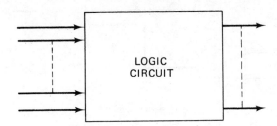

LOGIC
CIRCUIT

FIG. 6-30 Block Diagram, Multiple Output Circuit

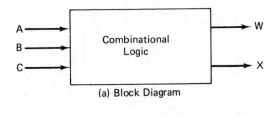

(a) Block Diagram

A	0000	1111
B	0011	0011
C	0101	0101
W	1110	0010
X	1101	1011

(b) Truth Table

FIG. 6-31 Multiple Output Minimization

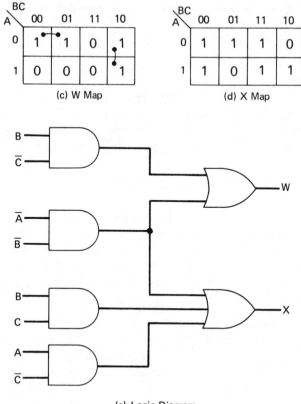

(e) Logic Diagram

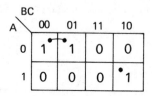

(f) WX Map

FIG. 6-31 (continued)

the W map are

$$W = \bar{A}\bar{B} + B\bar{C}$$

However, there are two possible ways to read the X map:

$$X = \bar{B}\bar{C} + \bar{A}C + AB$$

$$X = \bar{A}B + BC + A\bar{C}$$

But note that the W output has an $\bar{A}\bar{B}$ term in the expression. If we were to·choose the second X expression, we could share this $\bar{A}\bar{B}$ term between the W and the X outputs, as shown in Fig. 6-31(e), and further minimize the circuit. If we did not choose this X expression, the circuit would not be minimum. This sharing of gates forms the basis for multiple-output minimization.

It was fortunate that we saw that the $\bar{A}\bar{B}$ term could be shared between the W and X output expressions. There are several methods that allow us to see the sharing possibilities even more clearly without having to resort to as much intuition. One of these methods requires the construction of another map in which each output minterm is ANDed with the corresponding minterm of the second output and the results placed on this third map. Therefore, all the 1's on this third map represent minterms that are common between the two outputs. Figure 6-31(f) shows such a map for the previous illustration. By examining this map it becomes obvious that the $\bar{A}\bar{B}$ term is shared and can be used to minimize the circuit. The $AB\bar{C}$ term is also common. However, there is no way of reducing the input counts by its use.

Example 6-24: Solve the following multiple-output equations using mapping:

$$f_1 = \sum m(0, 1, 2, 4, 6, 7, 10, 14, 15)$$
$$f_2 = \sum m(3, 4, 5, 9, 10, 11, 14)$$

SOLUTION:

The maps for f_1 and f_2 are shown in Fig. 6-32(a) and (b). The terms common to f_1 and f_2 are

$$f_1 \cdot f_2 = \sum m(4, 10, 14)$$

This was also mapped, resulting in the possible dual use of the $AC\bar{D}$ term. Then, the f_1 and f_2 maps were minimized, making an effort to use this $AC\bar{D}$ term. The results were

$$f_1 = \bar{A}\bar{B}\bar{C} + \bar{A}D + BC + AC\bar{D}$$
$$f_2 = \bar{A}B\bar{C} + A\bar{B}D + \bar{B}CD + AC\bar{D}$$

The logic diagram is shown in Fig. 6-32(d).

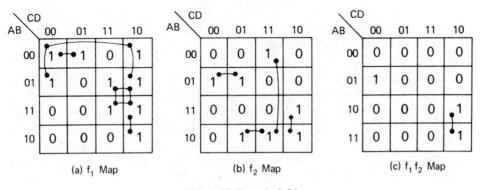

(a) f_1 Map   (b) f_2 Map   (c) $f_1 f_2$ Map

FIG. 6-32 Example 6-24

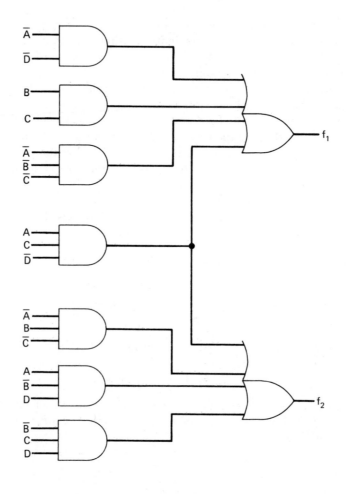

(d) Logic Diagram

FIG. 6-32 (continued)

Incompletely Specified Functions

Incompletely specified functions involving only two outputs can easily be solved using maps. The only difference between these and single-output procedures is the generation of the shared-term map. If this map is used to generate a shared gate, and a particular minterm covered by that gate must be a 1 for one of the functions, then it must be a 1 for the shared-term map. Therefore, the shared-term map must obey the rules listed in Table 6-3. For example, if minterm 13 of f_1 is a 1 (do care) and m_{13} of f_2 is a don't care, then m_{13} of the shared-term map must be a 1.

TABLE 6-3. Generating Minterms for a Shared-Term Map

f_1	f_2	$f_1 f_2$
0	0	0
0	1	0
1	0	0
0	X	0
X	0	0
1	X	1
X	1	1
1	1	1
X	X	X

Example 6-25: Design a multiple-output logic circuit using the following functions:

$$f_1 = \sum m(2, 3, 7, 10, 11, 14) + d(1, 5, 15)$$
$$f_2 = \sum m(0, 1, 4, 7, 13, 14) + d(5, 8, 15)$$

SOLUTION:

The two functions are first mapped [Fig. 6-33(a) and (b)]. Then, the shared-term map is generated according to the rules in Table 6-3; the map is shown in Fig. 6-33(c). Terms ABC, $\bar{A}\bar{C}D$, and $\bar{A}BD$ are candidates for shared terms. Since ABC must be generated for f_2, the same gate can be used for f_1. However, m_{13} of f_2 can only be covered using the four-square shown, so, rather than generate $\bar{A}BD$ for use only in f_1, four-square $\bar{A}D$ was chosen, since it is less expensive. The final functions are

$$f_1 = \bar{B}C + ABC + \bar{A}D$$
$$f_2 = \bar{A}\bar{C} + BD + ABC$$

The logic diagram is shown in Fig. 6-33(d).

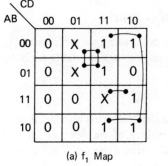

(a) f_1 Map

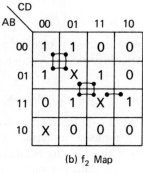

(b) f_2 Map

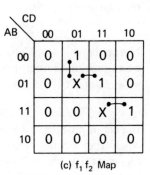

(c) $f_1 f_2$ Map

FIG. 6-33 Example 6-25

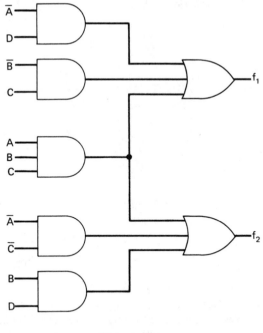

(d) Logic Diagram

FIG. 6-33 (continued)

6-10 VARIABLE MAPPING

The variable mapping technique is one that allows us to reduce a large mapping problem to one that uses just a small map. It is especially useful in those problems having a few isolated variables among more frequently used variables. Consider the equation:

$$Z = A\bar{B}\bar{C} + A\bar{B}C + \bar{A}BCR + ABC + AB\bar{C}$$

Normally, this would be a four-variable problem. However, using variable mapping we can make it into a three-variable problem. Assume this is a three-variable problem in A, B, and C. Thus, we have:

$$A = m_4 + m_5 + m_3(R) + m_7 + m_6$$

In such a problem we would put 1's on the map where the minterms appear and 0's where they do not. Each 1 represents a minterm. Therefore, if we put $1 \cdot R$, it will represent that variable multiplied by R. Figure 6-34(a) illustrates the map. We should next recognize that each 1 entered onto the map represents $R + \bar{R}$ [Fig. 6-34(b)]. Finally, we must cover each of the individual variables. This can be done by making a 2-square of the R in m_3 and the R in m_7 and making a 4-square of m_4, m_5, m_6 and

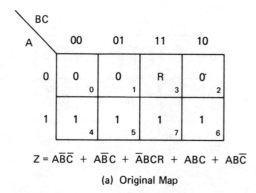

$$Z = A\bar{B}\bar{C} + A\bar{B}C + \bar{A}BCR + ABC + AB\bar{C}$$

(a) Original Map

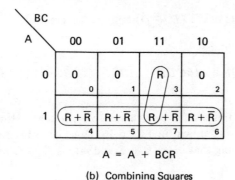

$$A = A + BCR$$

(b) Combining Squares

FIG. 6-34 Variable Mapping

m_7. The map is read as before except that squares with a single R term must be multiplied by R in the result. This yields:

$$Z = A + BCR$$

Note that the R in m_7 was covered twice. However, our objective is to cover all the variables at least once.

Example 6-26: Reduce by mapping:

$$M = \bar{A}\bar{B}C\bar{D} + \bar{A}\bar{B}CD + A\bar{B}C\bar{D} + A\bar{B}CD + \bar{A}BCD$$
$$+ \bar{A}B\bar{C}\bar{D} + AB\bar{C}D + AB\bar{C}\bar{D}$$

SOLUTION:

We can convert each of the four-variable minterms to three-variable minterms multiplied by a D term:

$$M = m_1\bar{D} + m_1D + m_5\bar{D} + m_5D + m_3D + m_2\bar{D} + m_6D + m_6\bar{D}$$

Now we can map each, including the D variable, on a 3-variable map [Fig. 6-35(a)]. Note that 1's are entered as $D + \bar{D}$. For example, $m_1 = m_1D + m_1\bar{D}$.

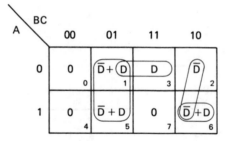

$$M = \bar{B}C + \bar{A}CD + B\bar{C}\bar{D} + AB\bar{C} \qquad \textbf{FIG. 6-35} \text{ Example 6-26}$$

The result is:

$$M = \bar{B}C + \bar{A}CD + B\bar{C}\bar{D} + AB\bar{C}$$

Note that all parts of a 1 $(D + \bar{D})$ must be covered.

Example 6-27: Reduce:

$$Z = \bar{A}\bar{B}CDE + \bar{A}B\bar{C}D\bar{E} + \bar{A}BCD + A\bar{B}C\bar{D}F + AB\bar{C}\bar{D} + ABC\bar{D}$$

SOLUTION:

Ordinarily, this would be a 6-variable problem. However, we can map it into 4 variables as shown in Fig. 6-36. Ones have been entered in m_7, m_{12}, and m_{14}, recognizing that m_7 is $E + \bar{E}$ and m_{14} is $F + \bar{F}$. Reading the map, we have:

$$Z = \bar{A}CDE + \bar{A}BD\bar{E} + AB\bar{D} + AC\bar{D}F$$

Note that this method greatly reduces the work required.

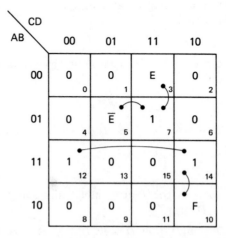

$$Z = \bar{A}CDE + \bar{A}BD\bar{E} + AB\bar{D} + AC\bar{D}F \qquad \textbf{FIG. 6-36} \text{ Example 6-27}$$

Incompletely Specified Functions

The procedure used for previous don't care problems has to be modified only slightly to accommodate variable mapping. Assume we had a three-variable problem, A, B, and C. Then an entry of 1 on the map represents $D + \bar{D}$. An entry of X (don't care) on the map represents $XD + X\bar{D}$, a don't care in both the D and the $\bar{D}$ minterms. For example, if m_6 is a don't care, we are saying that:

$$m_6 = AB\bar{C} = AB\bar{C}\bar{D} + AB\bar{C}D = m_6(\bar{D} + D)$$

Thus, whereas we would normally enter an X into square 6, we would enter $XD + X\bar{D}$. These don't cares may or may not be covered. They should be used to make 2^n squares when covering other do care squares. There is, however, one more case of a don't care: that of a particular minterm which is multiplied by the variable. Assume a 3-variable problem where $\bar{A}B\bar{C}\bar{D}$ is a don't care. We have:

$$\bar{A}B\bar{C}\bar{D} = m_2\bar{D}$$

Expressed as a don't care, square 2 would contain an $X\bar{D}$.

Example 6-28: Solve using 3-variable mapping:

$$W = Dm_2 + m_6 + Dm_5 + d(m_1 + \bar{D}m_7)$$

SOLUTION:

The map is shown in Fig. 6-37. Study each square carefully and observe how each was entered. The D in m_5 can only be covered by combining it with the XD of m_1. Similarly, the D of m_2 can only be combined with the D of m_6. This leaves the $\bar{D}$ of m_6 uncovered. We can combine it either with the D of m_6 or the $X\bar{D}$ of m_7. Both yield 3-term results. Reading the map:

$$W = \bar{B}CD + AB\bar{D} + B\bar{C}D$$

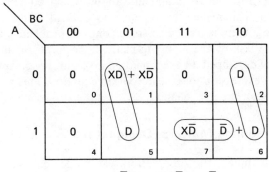

FIG. 6-37 Example 6-28 $W = \bar{B}CD + AB\bar{D} + B\bar{C}D$

Example 6-29: Reduce by mapping:

$$R = m_0 + (E + G)m_2 + \bar{E}m_5 + Gm_{10} + Fm_{13} + m_{14} + m_{15}$$
$$+ d(m_1 + Em_4 + \bar{F}m_8 + \bar{F}m_9)$$

SOLUTION:

This particular problem actually has 7 variables. However, we have factored to reduce it to 4 variables. These are plotted on Fig. 6-38. Note that, although the E of m_2 covers the E in m_0, the $\bar{E}$ of m_0 is still uncovered. Thus, it is combined with m_1. Reading the map:

$$R = \bar{A}\bar{B}\bar{D}E + \bar{A}\bar{B}\bar{C} + \bar{B}C\bar{D}G + ABDF + ABC$$

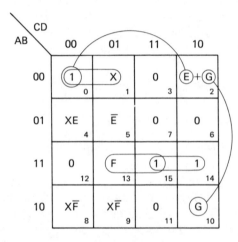

FIG. 6-38 Example 6-29

6-11 TABULAR MINIMIZATION

Although using maps is a fairly simple procedure, it has the disadvantage of requiring the designer to make judgments about which combinations form the minimum expression. This makes the system very difficult, if not impossible, to computerize. In addition, there is a need to solve complex problems of 7, 8, even 10 variables, an impossible task by the mapping method. To meet this need, W. V. Quine and E. J. McClusky developed an exact method involving an exhaustive tabular search for the minimum expression to solve a Boolean equation. This procedure is called the Quine-McClusky, or tabular, method.

Combining Minterm Designators

In mapping, the observation was made that any adjacent minterms can be reduced. Further, we observed that the reason they could be reduced is that they differed by only one literal. For example, $\bar{A}B\bar{C}\bar{D}$ and $\bar{A}BC\bar{D}$ can be reduced because only the C literal differs. But look closely at the minterm designators for the two terms. $\bar{A}B\bar{C}\bar{D} = m_4$ and $\bar{A}BC\bar{D} = m_6$. The 4 is a binary 0100 and the 6 a binary 0110. Note that the 2's place indicates precisely the same information that the literal represented, namely, that the C is the only literal that differs. Therefore, we can combine minterms whose binary designators differ only in one place. This forms the fundamental principle of the Quine-McClusky method of minimization.

But look again at the binary minterm designator. Note that a second condition is imposed by insisting that the binary designators differ only in one place: the total number of 1's in the two binary words can differ by only 1. Therefore, a word that has four 1's can only be combined with a word containing either three or five 1's.

Selecting Prime Implicants

Having presented the preliminaries, let us go through a problem in detail, solving it first by mapping and then by the Quine-McClusky method.

Example 6-30: Solve the following expression by mapping:

$$f = \sum m(0, 2, 3, 6, 7, 8, 9, 10, 13)$$

SOLUTION:

The map and its solution is shown in Fig. 6-39.

$$f = \bar{B}\bar{D} + \bar{A}C + A\bar{C}D$$

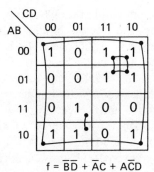

	CD	00	01	11	10
AB					
00		1	0	1	1
01		0	0	1	1
11		0	1	0	0
10		1	1	0	1

FIG. 6-39 Example 6-30 $f = \bar{B}\bar{D} + \bar{A}C + A\bar{C}D$

Next, we shall solve the problem using the Quine-McClusky method.

First, list the binary forms of all the minterm designators as shown in Table 6-4(a). Then arrange these binary words in the order of the number of 1's contained in each word, drawing a horizontal line between each number-of-1's category [Table 6-4(b)].

Next, compare each binary number with every term in the next higher category, and if they differ by only one position, put a check mark beside each of the two terms and then copy the term down with an X in the position that they differed. Also write

TABLE 6-4. Selecting Prime Implicants

Minterm	Binary Desig.		Minterm	Binary Desig.		Number of 1's
m_0	0000		0	0000 ✓		0
m_2	0010		2	0010 ✓		1
m_3	0011		8	1000 ✓		
m_6	0110		3	0011 ✓		
m_7	0111		6	0110 ✓		2
m_8	1000		9	1001 ✓		
m_9	1001		10	1010 ✓		
m_{10}	1010		7	0111 ✓		3
m_{13}	1101		13	1101 ✓		

(a) Binary Designators   (b) Number-of-1's Categories

0, 2	00X0 ✓	
0, 8	X000 ✓	
2, 3	001X ✓	
2, 6	0X10 ✓	
2, 10	X010 ✓	
8, 9	100X	
8, 10	10X0 ✓	
3, 7	0X11 ✓	
6, 7	011X ✓	
9, 13	1X01	

(c) First-Level Implicants

0 2, 8, 10	X0X0	
2, 3, 6, 7	0X1X	

(d) Second-Level Implicants

the terms the X'ed term covers beside the X'ed term. For example, 0000 and 0010 differ only in the 2's column. Put a check beside 0000 and 0010, copy 00X0 down, and copy the terms it covers, 0 and 2 [Table 6-4(c)]. This term, 00X0, is called an implicant, for it implies that $\bar{A}\bar{B}\bar{D}$ could be a term in the final expression, covering minterms $\bar{A}\bar{B}\bar{C}\bar{D}$ and $\bar{A}\bar{B}C\bar{D}$. Next, 0000 and 1000 combine, so put a check beside 1000 (0000 already has a check) and write implicant X000 below implicant 00X0, indicating it covers minterms 0 and 8. We have now examined every category 0-1 combination, so draw a horizontal line below X000. Next examine category 1-2 combinations, yielding implicants 001X, 0X10, X010, 100X, and 10X0. Copy these down along with the terms covered and draw a horizontal line below 10X0. The procedure is repeated for the 2-3 category.

Thus far, we have found 10 possible two-squares. However, we now have to examine for possible four-squares. Therefore, compare each binary designator in the first-level-implicant table with each member of the next highest category of implicants. They must differ by only one 1 and both must contain X's in the same column. For example, 00X0 does not combine with 001X, 0X10, X010, or 100X. It does combine with 10X0, yielding X0X0. Similarly, X000 combines only with X010, yielding implicant X0X0, a four-square. Continue the process as shown.

All those terms that remain unchecked are called prime implicants, for they cannot be reduced any further (made a part of a larger 2^n-square). They represent all the possible two- and four-squares that can be obtained from a Karnaugh map. Not all of them will be used, but the final expression will be a sum of one or more prime implicants.

Selecting Output Terms

Having found the possible terms that could appear in the output expression, we must now select the terms necessary for a minimal expression. First, draw a table listing all the minterms along the top and all the prime implicants along the side in descending order of squares covered [Table 6-5(a)]. This example has two four-square implicants and two two-square implicants. Draw a line to separate the two-square from the four-square implicants. Put dots under the terms covered by each implicant.

Examine minterm 0. Since there is only one prime implicant covering this minterm, $X0X0$ is said to be an essential prime implicant. Therefore, it must be included in the output expression, as indicated by the asterisk next to $X0X0$ [Table 6-5(b)]. But if we include four-square $X0X0$, we shall also be covering minterms 0, 2, 8, and 10, so place a check at the foot of each of these minterm columns.

Next, examine minterm 3 and find essential prime implicant $0X1X$, which also covers m_2 (already checked), m_6, and m_7. Place a check under all these columns, indicating these terms are now covered [Table 6-5(c)]. This leaves m_9 and m_{13} uncovered. But m_9 can be covered by either of two prime implicants, so we cannot make a decision yet. However, m_{13} can only be covered by $1X01$, so it must be included. As it happens, it also covers m_9, [Table 6-5(d)].

From this tabular reduction, these prime implicants, $X0X0$, $0X1X$, and $1X01$, must now be converted to a Boolean expression. To do this, substitute literals for the 1's, negated literals for the 0's, and omit the X's, giving $\bar{B}\bar{D}$ for $X0X0$, $\bar{A}C$ for $0X1X$, and $A\bar{C}D$ for $1X01$. The final expression is

$$f = \bar{B}\bar{D} + \bar{A}C + A\bar{C}D$$

the same as was obtained by mapping.

6-12 SUMMARY

Combinational logic is that logic that, neglecting transit time, responds immediately to a combination of inputs. Each combination of these inputs is called a minterm and can be plotted on a truth table or a Karnaugh map. Maps can be reduced by combining minterms into 2^n-squares, and then reading the results in either of two forms: sum-of-products or product-of-sums. These results are then compared, using total gate inputs as the criterion. Incompletely specified functions are those for which one or more minterms can be either logic level. Multiple-output reduction provides opportunity for reducing cost through sharing of gates between output functions. Tabular minimization allows a computer to perform Boolean reduction.

TABLE 6-5. Prime Implicant Tabular Reduction

		0	2	3	6	7	8	9	10	13
X0X0		•	•				•		•	
0X1X			•	•	•	•				
100X							•	•		
1X01								•		•

(a) Original Table

		0	2	3	6	7	8	9	10	13
X0X0	*	•	•				•		•	
0X1X			•	•	•	•				
100X							•	•		
1X01								•		•
		√	√				√		√	

(b) First Step

		0	2	3	6	7	8	9	10	13
X0X0	*	•	•				•		•	
0X1X	*		•	•	•	•				
100X							•	•		
1X01								•		•
		√	√	√	√	√	√		√	

(c) Second Step

		0	2	3	6	7	8	9	10	13
X0X0	*	•	•				•		•	
0X1X	*		•	•	•	•				
100X							•	•		
1X01	*							•		•
		√	√	√	√	√	√	√	√	√

(d) Final Step

6-13 PROBLEMS

6-1. Convert $\bar{A}B + C$ to minterms.

6-2. Convert $A + B + C$ to minterms.

6-3. Convert $AB + \bar{C}D + A\bar{B}C$ to minterms.

6-4. Convert $AB + AC + A\bar{D}$ to minterms.

6-5. Find the minterm designators for the following minterms:
  (a) $A\bar{B}C$ (b) $A\bar{B}C\bar{D}$ (c) $AB\bar{C}\bar{D}E$

6-6. Find the minterms designators for the following minterms:
  (a) $\bar{A}\bar{B}\bar{C}$ (b) $A\bar{B}C\bar{D}$ (c) $AB\bar{C}\bar{D}E$

6-7. Place on a truth table and map the function $C = A\bar{B} + AB + \bar{A}B$.

6-8. Place on a truth table and map the function $X = \bar{A}BC + \bar{A}\bar{B}\bar{C}$.

6-9. Place on a truth table and map the function

$$R = A\bar{B}\bar{C}D + A\bar{B}C\bar{D} + \bar{A}BCD$$

6-10. Place on a truth table and map the function

$$S = A\bar{B}C\bar{D}E + \bar{A}\bar{B}\bar{C}DE + ABC\bar{D}\bar{E} + \bar{A}BC\bar{D}\bar{E}$$

6-11. Place on a truth table and map the equation

$$f = \sum m(0, 1, 3, 4, 7, 10, 20, 28, 30)$$

6-12. Place on a truth table and map the function

$$X = \bar{A}BC\bar{D}EF + A\bar{B}C\bar{D}\bar{E}F + AB\bar{C}\bar{D}EF + ABC\bar{D}EF + \bar{A}B\bar{C}DEF + ABC\bar{D}\bar{E}F$$

6-13. Reduce by mapping the implement in **SOP NAND** logic:

$$M = \sum m(0, 1, 3, 4, 5, 7, 10, 13, 14, 15)$$

6-14. Reduce by mapping and implement in **SOP NAND** logic:

$$N = \sum m(0, 2, 3, 6, 7, 8, 10, 11, 12, 15)$$

6-15. Reduce by mapping and implement in **SOP NOR** logic:

$$P = \sum m(1, 2, 3, 4, 6, 7, 10, 11, 13, 14)$$

6-16. Reduce by mapping and implement in **SOP NOR** logic:

$$Q = \sum m(2, 3, 4, 6, 7, 10, 13, 14)$$

6-17. How many gate inputs are required for the following equations:
  (a) $W = A\bar{B}C + \bar{A}B + A\bar{C}D$
  (b) $X = R\bar{S} + R\bar{T}\bar{U}V + RSTUV + TUV$
  (c) $Y = (A + B + C)(A + \bar{B} + D)(\bar{A} + \bar{C})(A + \bar{B} + \bar{C} + D)$

6-18. How many gate inputs are required for the following equations:
  (a) $L = A\bar{B} + A\bar{C}D + \bar{B}CD$
  (b) $M = EFG\bar{H}I + GH + F\bar{H}\bar{I} + \bar{E}I + \bar{E}\bar{F}$
  (c) $N = (\bar{J} + K)(L + \bar{M} + N + \bar{P})(\bar{J} + L + \bar{M} + \bar{N})(\bar{M} + N)$

6-19. Reduce the following equation to simplest **SOP** form using mapping:

$$f = \sum m(0, 2, 3, 10, 11, 12, 13, 16, 17, 18, 19, 20, 21, 26, 27)$$

6-20. Reduce the following equation to simplest SOP form using mapping:

$$f = \sum m(4, 6, 10, 11, 14, 15, 20, 22, 26, 27, 30, 31, 36, 38, 39, 52, 54, 55, 56, 57, 60, 61)$$

6-21. Design an SOP circuit that will output a 1 anytime the Gray codes 5 through 12 appear at the inputs and a 0 for all other cases.

6-22. Design a four-input SOP circuit that will output a 1 anytime the input is the binary equivalent of 0, 1, 5, 9, 11, 13, or 15 decimal.

6-23. Design an SOP circuit that will output a 1 anytime a letter within my name, William Henry Gothmann, appears at the input in five-level teletypewriter code and a 0 for all other characters.

6-24. Design an SOP circuit that will output a 1 anytime a letter within my name, William Henry Gothmann, appears at the input in ASCII code. Assume that only characters with most significant bits of 100 and 101 will ever appear at the inputs.

6-25. Reduce to simplest POS form and implement in **NAND** logic:

$$E = \sum m(0, 1, 3, 4, 5, 6, 7, 13, 15)$$

6-26. Reduce to simplest POS form and implement in **NOR** logic:

$$S = \sum m(0, 2, 4, 6, 10, 11, 12, 14, 15)$$

6-27. Reduce to simplest form (POS or SOP) and implement in **NOR** logic:

$$T = \sum m(0, 1, 3, 4, 6, 7, 10, 14)$$

6-28. Reduce to simplest form (POS or SOP) and implement in **NAND** logic:

$$U = \sum m(1, 2, 3, 6, 10, 13, 14, 15)$$

6-29. Find both the reduced SOP and POS forms for the equation

$$L = \prod M(0, 1, 2, 3, 4, 6, 10, 11, 13)$$

6-30. Find both the reduced SOP and POS forms of the equation

$$N = \prod M(0, 1, 2, 3, 8, 9, 10, 11, 14, 15, 20, 21, 22, 23, 24, 25)$$

6-31. A circuit receives four-bit 8421 BCD code. Design the minimum circuit to detect the decimal numbers 1, 2, 3, 6, 7, and 8. Implement in **NAND** logic.

6-32. A circuit receives four-bit 5211 BCD code. Design the minimum circuit to detect the decimal numbers 0, 2, 4, 6, and 8. Implement in **NAND** logic.

6-33. A circuit receives only the characters A through Z in ASCII code from the transmitter. Design a minimum circuit such that it will detect any of the letters used within the last five words of this sentence.

6-34. A circuit consists of two control lines, C_1 and C_2, and three data lines, X, Y, and Z. Design the minimum POS circuit that will perform the following functions:

Control Lines	Response
00	Transmit a 1 if any data bits are 1
01	Transmit a 1 if any data bits are 0
10	Transmit a 1 if two or more data bits are 1
11	Will never occur

6-35. Minimize and implement the following multiple-output functions in SOP form:

$$f_1 = \sum m(1, 2, 3, 6, 8, 12, 14, 15)$$
$$f_2 = \sum m(1, 2, 3, 5, 6, 7, 8, 12, 13)$$

6-36. Minimize the following multiple-output functions using SOP:

$$f_1 = \sum m(0, 1, 2, 4, 6, 7, 10)$$
$$f_2 = \sum m(2, 4, 10, 12, 14)$$

6-37. Minimize the following multiple-output functions:

$$f_1 = \sum m(0, 2, 6, 10, 11, 12, 13) + d(3, 4, 5, 14, 15)$$
$$f_2 = \sum m(1, 2, 6, 7, 8, 13, 14, 15) + d(3, 5, 12)$$

6-38. A circuit receives only valid 5211 or 8421 BCD information and provides two output lines, X and Y. Design the circuit such that X will provide an output anytime a valid 8421 BCD code appears at the input and Y will provide an output anytime a valid 5211 BCD code appears at the input.

Solve using a 3-variable map:

6-39. $N = \bar{A}\bar{B}\bar{C} + \bar{A}\bar{B}CD + A\bar{B}\bar{C}\bar{D} + ABC + AB\bar{C}$

6-40. $M = \bar{B}\bar{C} + AB\bar{C} + \bar{A}BC\bar{D} + \bar{A}B\bar{C}D$

6-41. $Y = Dm_3 + m_2 + m_4 + d(m_0 + Dm_1)$

6-42. $Z = Dm_3 + m_5 + \bar{D}m_7 + d(Dm_2 + m_4)$

Solve using a 4-variable map:

6-43. $Y = Em_0 + m_4 + m_{10} + Fm_{11} + \bar{E}m_{12} + m_{14} + Fm_{15}$

6-44. $S = m_1 + m_5 + Gm_6 + \bar{F}m_8 + \bar{F}m_9 + Fm_{11} + m_{12} + m_{13} + m_{14} + m_{15}$

6-45. $T = m_3 + (\bar{F} + G)m_6 + m_7 + d(m_1 + m_5 + m_8 + m_{13} + Gm_{14})$

6-46. $L = Hm_1 + Em_6 + m_{12} + (F + \bar{G})m_{13} + d(Hm_0 + m_2 + (\bar{E} + H)m_3 + Em_{10} + m_{14})$

SEQUENTIAL CIRCUITS

7

7-1 INTRODUCTION

We deal with sequential devices every day without calling them as such. For example, we all have operated a combination lock. This device requires the operator to provide the correct numbers, but they must be in the correct time sequence. Therefore, it is not only a combinational problem but also a sequential problem.

Division is another example. Not only must addition, subtraction, and multiplication be performed to solve the problem, but these operations must be performed in the proper sequence.

A computer executes instructions in such a time sequence, performing the first, then the second, then the third, and so forth. If it did not, we might find our dental bill being sent by the grocer and our bank account overdrawn by the next door neighbor. Thus, time sequence is important to the computer.

In digital electronics, there are two types of sequential circuits: clocked and unclocked, referred to as *synchronous* and *asynchronous*. In synchronous logic, there is a master oscillator that provides regular timing pulses. The only time that events are permitted to occur is during one of these timing pulses. Except for propagation delays, the circuit is inert at other times. A magnetic tape reader is an example of this type of circuit. Reading only occurs while the tape is under the read head.

In asynchronous systems, events occur after the previous event is completed; there is no need to wait for a timing pulse. An example of this is a touch tone telephone system; here, the user punches the next number as soon as he's finished punching the previous one.

In this chapter we shall examine both types of logic systems, detailing their design techniques, advantages, and disadvantages.

7-2 FLIP-FLOPS

We store information in a variety of ways. For example, the words you are now read-ing have been stored there by the printer; if you were to lay this book down (don't do it, or you will never know what will happen) and 5 years from now pick it up, the information would still be there, God and the termites willing. We store informa-tion in a light switch; we turn it to the on state, and sure enough, it remains in the on state until the plug is pulled or the switch is thrown. We also store information in punched cards, paper tape, magnetic tape, phonograph records, and a multitude of other ways. In each of these media, if we were to flee the scene and later return, the information would be preserved.

To have a sequential system, we must know what has happened in the past. Therefore, we must have storage devices to retain this information until we are ready to use it. The basic unit for this storage is the *flip-flop* (abbreviated FF). There are, however, several varieties of the critter, which we shall discuss.

RS Flip-flop

The simplest type of storage device is the reset-set (RS) flip-flop (FF). It can be formed out of any two **NAND** or **NOR** gates you happen to have lying around your house. Figure 7-1(a) illustrates the **NAND** connection. It has two outputs, Q and $\bar{Q}$ (this is standard nomenclature for any FF). Assume that Q is a logical 1 and that both switches are open. Then both inputs to A_2 are high, resulting in $\bar{Q}$ being low. However, A_1 has one low and one high input; its output must therefore be high, as we assumed. All this has shown is that $Q = 1, \bar{Q} = 0$, is a stable state; the device will remain in that state until told to do otherwise.

Let us now make the opposite assumption, that Q is a 0, low. Then A_2 has a low input, making its output, $\bar{Q}$, high. But if A_1 has two highs on its inputs, its output, Q, must be low. Again, this is a stable state. We have, therefore, a device capable of two stable states. If we arbitrarily call the state when Q is a 1 the one state, we can call the state when Q is a 0 the zero state. This device is, therefore, capable of storing one bit of information, a 0 or a 1.

But why would I put switches on a diagram without using them? Lest you be dis-appointed, let us examine what happens when S_1 closes. This puts a low on A_1, and, regardless of its other input, Q will go high. This will place two highs on A_2, causing $\bar{Q}$ to go low. If we were now to release (open) the switch, the device would remain in this one state. Thus, regardless of what state it was in, the FF is now in the one state.

Now try closing S_2. This puts a low into A_2, making $\bar{Q}$ high and Q low. There-fore, we can say that S_1 enters a 1 into the FF (it puts it into the one state), and S_2 enters a 0 into the FF. Remember, the state of the FF is determined by the Q output. If Q is a 1, the FF is in the one state. If it is a 0, then the FF is in the zero state.

The **NOR** gate RS FF functions in a similar manner [Fig. 7-1(b)], except that

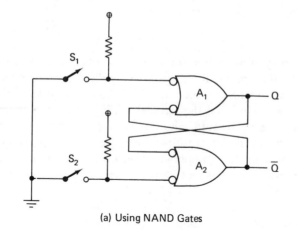

(a) Using NAND Gates

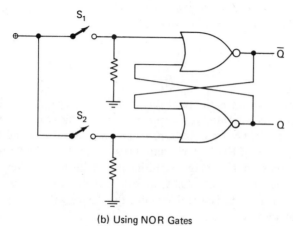

(b) Using NOR Gates

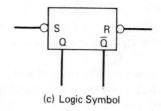

(c) Logic Symbol   **FIG. 7-1** The *RS* Flip-flop

its inputs are normally grounded and switched to the high state by closing the switches.

The logic symbol for the *RS* FF is shown in Fig. 7-1(c). Grounding the *S* lead sets the FF to a 1; grounding the *R* lead resets it to a 0. Logic rules do not permit the *S* and *R* lines to be activated simultaneously. The set and reset lines are called preset and clear by some manufacturers.

Type *T* Flip-flop

The type T FF, (Fig. 7-2) changes state (toggles) each time a pulse is received on its T input. Assume a square wave is applied to the T input. The Q and $\bar{Q}$ outputs would be as shown in the figure. Note that its output frequency is one-half its input frequency. It can, therefore, be used for both a counter and a frequency divider.

Set and reset inputs can be added to the type T. They always take priority over any other input. If the T input received a square wave and the S input were activated, the FF would go to the one state and stay there.

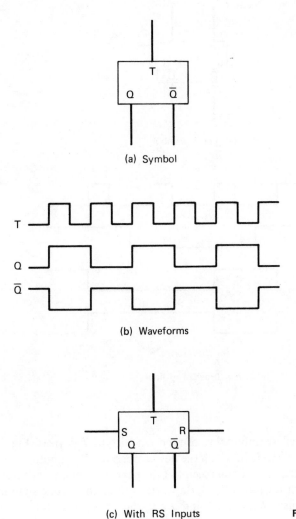

(a) Symbol

(b) Waveforms

(c) With RS Inputs

FIG. 7-2 The Type *T* Flip-flop

Type D Flip-flop

The type D (delay) FF is used for storing infromation. It has two inputs, a clock (called CK, CL, C, or T, depending on the manufacturer) and a D input. Information appearing on the D input will be transferred to the FF outputs upon receipt of the next clock pulse. Figure 7-3 illustrates its logic diagram and waveform examples. Note that it effectively stores the information on the D line at the time of the clock pulse.

Set and reset inputs may be added to the unit, in which case they take priority over any other inputs.

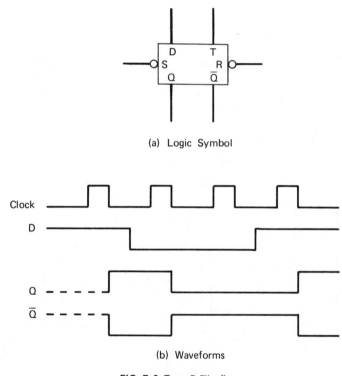

(a) Logic Symbol

(b) Waveforms

FIG. 7-3 Type D Flip-flop

Type JK Flip-flop

The type JK flip-flop is a combination of type D and type T devices (Fig. 7-4). The J and K leads control the mode in which the device operates. With both the J and K 0's, the FF will remain unchanged upon receipt of the next clock pulse; the information remains unaltered. With J a 1 and K a 0, the device will go to the one state upon receipt of the next clock pulse. With K a 1 and J a 0, the device will go to the zero

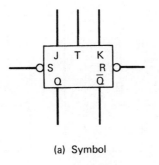

(a) Symbol

J K	Response
0 0	Unchanged
0 1	0
1 0	1
1 1	Flip

(b) Effect of JK Inputs

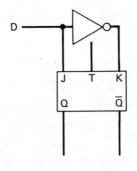

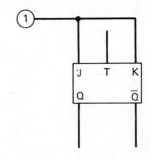

FIG. 7-4 Type *JK* Flip-flop   (c) Type D Connection   (d) Type T Connection

state upon receipt of the next clock pulse. With both *J* and *K* 1's, the next clock pulse will cause the device to flip to the opposite state in typical *T* fashion. This operation is summarized in Fig. 7-4(b).

By connecting the *JK* FF as shown in Fig. 7-4(c), it can be used as a type *D*; by connecting it as shown in Fig. 7-4(d), it becomes a type *T*. This is why it is called a universal FF.

As in the previous FF's, set and reset inputs can be added to the device, in which case they take priority over any other input.

In the next section we shall describe how each of these FF's is used in counter designs.

7-3 RIPPLE COUNTERS

A ripple counter is a counter that uses type *T* FF's to perform a counting function, where each *T* lead is connected to an output of the previous stage. The actual hardware used is usually a *JK* FF with the *J* and *K* leads wired high [Fig. 7-4(d)]. Figure 7-5 illustrates such a counter that counts through four different states. Each FF is

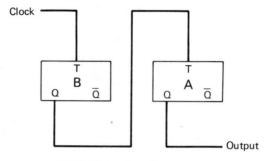

(a) Divide by Four Counter

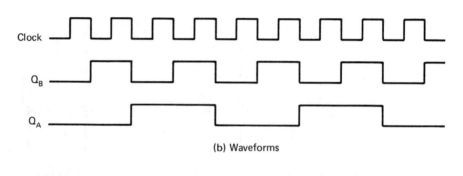

(b) Waveforms

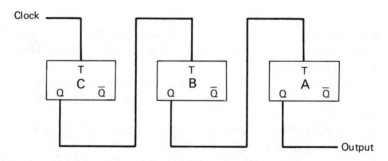

(c) Modulo 8 Counter

FIG. 7-5 Ripple Counter

designed to flip when its T lead from high to low. Therefore, Q_B counts by 2's and Q_A by 4's. If the frequency of the clock pulse is 4 kilohertz (kHz), the frequency of Q_B would be 2 kHz and Q_A 1 kHz. Therefore, the circuit is also called a divide-by-4 circuit.

The circuit goes through four different sequences before it arrives in the same sequence in which it started:

B	A
0	0
1	0
0	1
1	1
0	0
etc.	

Note that FF B flips faster than FF A. Therefore, if we were to call B the least significant bit of the counter, the sequence would be

A	B	Decimal
0	0	0
0	1	1
1	0	2
1	1	3
0	0	0
etc.		

and, as can be seen, it becomes a plain binary counter. Each of these counts is said to be a *state*. Therefore, this counter goes through states 0, 1, 2, and 3 before returning to 0. Again, note that the least significant bit of any counter is that bit that changes most often.

This ripple counter goes through four states and then starts over again. For this reason, it is called a modulo-4 counter. But what if a modulo-8 counter is needed? One more flip-flop will do the job [Fig. 7-5(c)]. Note that a two-stage counter has 2^2 or 4 unique states and a three-stage counter 2^3 or 8 unique states. An n-stage counter will have 2^n unique states.

Shortened Modulus

The next step is to design a modulo-3 (not a power of 2) ripple counter. To do this, we must first draw a state diagram. This gives a picture of what states the circuit will go through [Fig. 7-6(a)]. Note that this circuit will have three states: 0, 1, and 2. This type of design will require FF's that have a reset input. The design technique is to detect the proper state using combinational logic and feed a reset pulse to the counter, resetting it to state 0 [Fig. 7-6(b)]. This reset pulse is very narrow and exists only until the FF's have been reset.

The next question we must answer is how many FF's to use. We are counting to state 2. One FF can count to state 1 and two FF's to state 3. Therefore, we need two FF's so we can have a state 2. Having decided this, the truth table for the reset pulse is constructed. We shall feed the reset pulse to the counter during the state following the last desired state in the sequence. Therefore, when it reaches state 3, the reset pulse will go true, immediately resetting the circuit to state 0, which will then remove the

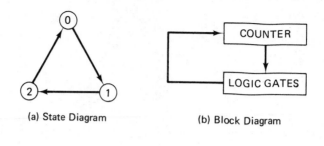

(a) State Diagram   (b) Block Diagram

A   0011
B   0101
─────────
Reset 0001

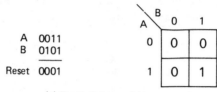

(c) Truth Table and Map

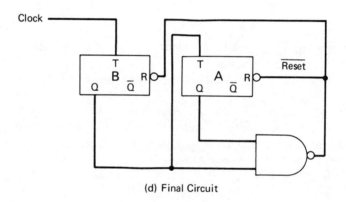

(d) Final Circuit

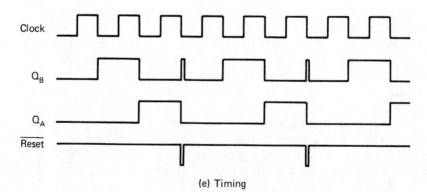

(e) Timing

FIG. 7-6 Modulo-3 Counter

reset pulse [Fig. 7-6(c)]. Notice that the counter will only be in state 3 for just an instant, whereas it will be in the other states for a full clock pulse period. The problem is then mapped, resulting in $R = AB$ for the pulse.

Finally, the circuit is constructed, taking note that B is the LSB and A the MSB [Fig. 7-6(d)]. The reset pulse is a logical **AND** of A (the Q output of A) and B (the Q output of B).

Example 7-1: Design a modulo-14 ripple counter.

SOLUTION:

The first step is to draw a state diagram with states 0 through 13 (14 total states) [Fig. 7-7(a)]. This will require four FF's, since three will give only 2^3 or

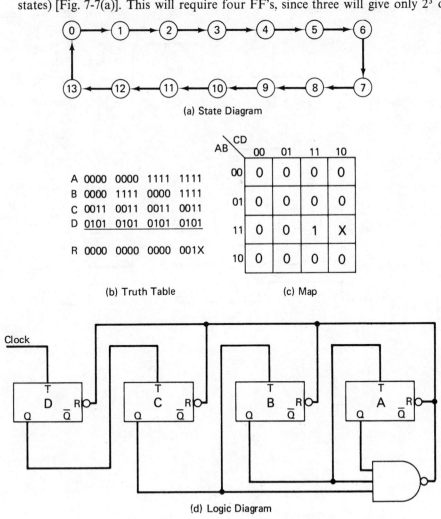

(a) State Diagram

| | | A 0000 0000 1111 1111 |
| B 0000 1111 0000 1111 |
| C 0011 0011 0011 0011 |
| D 0101 0101 0101 0101 |

R 0000 0000 0000 001X

AB\CD	00	01	11	10
00	0	0	0	0
01	0	0	0	0
11	0	0	1	X
10	0	0	0	0

(b) Truth Table

(c) Map

(d) Logic Diagram

FIG. 7-7 Example 7-1, Modulo-14 Counter

8 states and four will provide 2^4 or 16 states. Next, the truth table is constructed for the reset pulse [Fig. 7-7(b)]. Note that state 14 will be the reset state and that state 15 is a don't care, for it will never be encountered. This is then mapped, resulting in $R = ABC$. Finally, the logic diagram constructed [Fig. 7-7(d)].

Down Counters

One problem that occurs constantly in ripple counters is determining whether a circuit is counting up or counting down. In the previous examples, we assumed that all FF's toggle (change states) on the positive to negative transition of the toggle pulse. However, some do and some do not. Let us redraw our divide-by-4 circuit assuming the FF's toggle on the negative to positive transition (Fig. 7-8). This results in a count

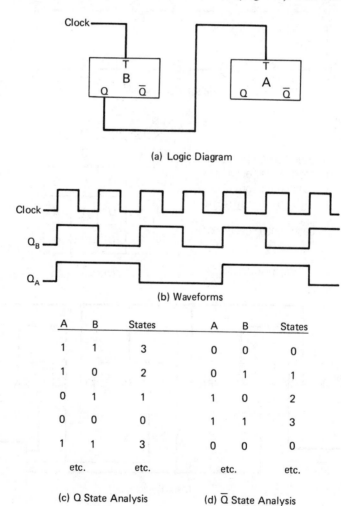

(a) Logic Diagram

(b) Waveforms

A	B	States	A	B	States
1	1	3	0	0	0
1	0	2	0	1	1
0	1	1	1	0	2
0	0	0	1	1	3
1	1	3	0	0	0
etc.		etc.	etc.		etc.

(c) Q State Analysis　　　(d) $\overline{Q}$ State Analysis

FIG. 7-8 Negative to Positive Toggle Flip-flop

down counter, rather than a count up counter. However, take a close look at the $\bar{Q}$ outputs of the FF's. If we were to take our outputs off these, the state analysis shows we would have a count up counter again [Fig. 7-8(d)].

Consider the logic diagram shown in Fig. 7-9. Assume those are the same FF's as used in Fig. 7-6 and Example 7-1 (they toggle on positive to negative transitions). The only thing that has been changed is that T of FF A receives its pulse from $\bar{Q}$ of B rather than Q of B. Note that, again, we have a down counter. Thus, there are three factors affecting whether a ripple counter counts up or down:

A. Whether the FF's toggle on a positive to negative or negative to positive transition.
B. Whether the Q or $\bar{Q}$ is used to feed the next T input.
C. Whether the output is taken off the Q or $\bar{Q}$ outputs.

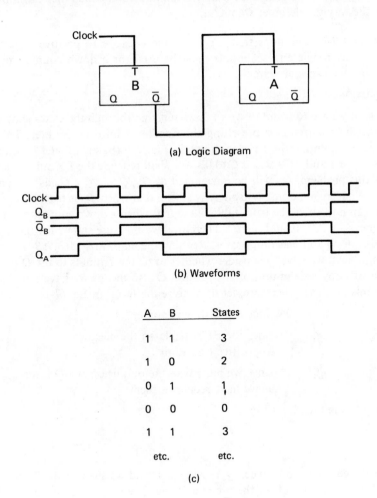

(a) Logic Diagram

(b) Waveforms

A	B	States
1	1	3
1	0	2
0	1	1
0	0	0
1	1	3
etc.		etc.

(c)

FIG. 7-9 Toggling Off $\bar{Q}$ Output

These factors can be exclusively **OR**ed in a Boolean expression:

$$U_p = P_n \oplus Q_t \oplus Q_1 \qquad (7\text{-}1)$$

where U_p represents an up counter, P_n means a positive to negative transition, Q_t represents toggling off the Q outputs, and Q_1 signifies taking the outputs off the Q output. If all these are true, then

$$U_p = 1 \oplus 1 \oplus 1 = 1$$

and the device will be an up counter. If, however, $U_p = 0$, then the result is a down counter.

There is one more "hitch in the git-along" with ripple counters. Many type T FF's have only a set input, and no reset. However, this is easily overcome by calling the set lead the reset lead, calling the Q lead the $\bar{Q}$ lead, and calling the $\bar{Q}$ lead the Q lead. It really makes no difference what the manufacturer called them, as long as they function the way the designer wants.

> **Example 7-2:** Flip-flops that toggle on the negative to positive transition and have only toggle and set inputs must be used for a down counter, modulo 5. Design the necessary circuit.

SOLUTION:

If we are to count down to 0, we must go through the states shown in Fig. 7-10. But this requires presetting the number 4 into the counter. This can be done by orienting the FF's as shown in Fig. 7-10(b) such that FF A, the MSB, is set to a 1 and FF's B and C to 0's. We shall redefine the FF outputs such that, for this problem, Q_A will be interpreted as $\bar{Q}_A$, $\bar{Q}_A$ as Q_A, Q_B as $\bar{Q}_B$, and $\bar{Q}_B$ as Q_B. What we are doing is arbitrarily saying that when $\bar{Q}_C$ is true, C contains a 1. We can now draw the truth table. Note that when a down counter counts down, it goes 4, 3, 2, 1, 0, and then 7. Therefore, we shall use state 7 for the reset pulse and states 6 and 5 as don't cares. Reading the map, $R = AC$ (AB is also a solution). Next, we have to consider whether to tie the T inputs to the $\bar{Q}$ or Q leads. Our arbitray redefinition of FF's B and C also means we have redefined them for this purpose. Therefore, let us analyze the terms in Eq. (7-1):

$U_p = 0$ since it is a down counter

$P_n = 0$ since the FF's toggle on the negative to positive transition

$Q_1 = 1$ since we must take the output off the Q leads as we have redefined them

Substituting into Eq. (7-1),

$$U_p = P_n \oplus Q_t \oplus Q_1$$
$$0 = 0 \oplus Q_t \oplus 1$$

For the equation to be true, Q_t must be a 1, and we shall tie the T inputs of the FF's to the Q outputs of the previous stage. The logic diagram for the device is

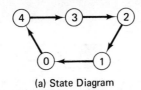

(a) State Diagram

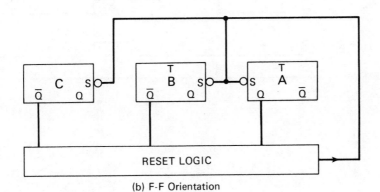

(b) F-F Orientation

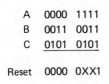

A   0000   1111
B   0011   0011
C   0101   0101

Reset   0000   0XX1

(c) Truth Table

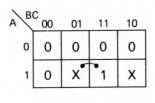

(d) Map

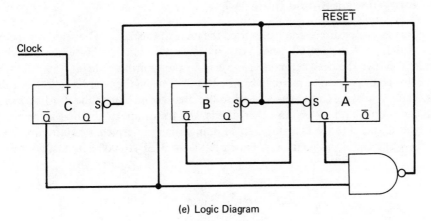

(e) Logic Diagram

FIG. 7-10 Example 7-2

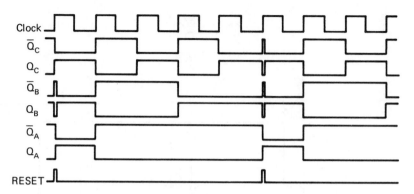

(f) Timing Diagram

Q_A	$\overline{Q}_B$	$\overline{Q}_C$	State
1	0	0	4
0	1	1	3
0	1	0	2
0	0	1	1
0	0	0	0
1	1	1	Reset to 4

(g) State Analysis

FIG. 7-10 (continued)

shown in Fig. 7-10(e), the timing in (f), and its state analysis in (g). As can be seen, it satisfies the specifications.

But What's a Ripple Counter?

This extensive discussion has skirted the issue of the day, "What is a ripple counter?" Although it is generally the least expensive of counters, the ripple of a ripple counter can give the designer cause for a trip to the aspirin bottle.

This counter is called a ripple counter because when the counter goes from 1111 to 0000 the first stage causes the second to flip, the second causes the third to flip, the third the fourth, and so on. In other words, the transition of the first stage ripples through the stages to the last stage. In so doing, many intermediate states are briefly entered. For example, to go from 1111 (the left is the MSB) to 0000, it must go through

1111	State 15
1110	State 14
1100	State 12
1000	State 8
0000	State 0

If the designer has a gate that will **AND** during state 12, a brief spike will be seen at the gate output every time the counter goes from state 15 to state 0. This is also true when it goes from state 13 to 14: 1101, then 1100, then 1110.

The ripple counter is, then, a very inexpensive and a very dirty counter, insofar as unwanted spikes are concerned. The design procedures that follow (type *T*, type *D*, and type *JK*) overcome this serious limitation.

7-4 TYPE *T* DESIGN

To distinguish this design procedure from that in Sec. 7-3, we shall call this a type *T* design as opposed to the ripple counter. Both, in fact, use type *T* FF's, usually *JK* FF's in a type *T* configuration. However, the ripple counter connects the *T* lead to an output of the previous stage, whereas the type *T* design connects all *T* leads to the clock pulse (Fig. 7-11). For this reason, this type of counter is often called a parallel counter. The *J* and *K* leads are each wired as shown and controlled by combinational logic. For simplicity, we shall refer to these leads as G_A, G_B, and G_C, with each *G* lead representing the output of a logic gate. By the definition of a *JK* FF, if a *G* lead is low and a clock pulse received, the FF will not flip. If a *G* lead is high and a clock pulse received, it will flip. Therefore, we can control what state we wish the counter to enter

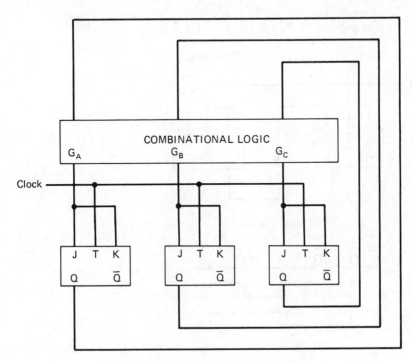

FIG. 7-11 Type *T* Design, Block Diagram

by controlling the G logic levels. Note that all stages of this type of counter enter the next state at precisely the same time: when the clock pulse makes its transition from positive to negative, for example. It is, therefore, a purely synchronous counter.

With this introduction, let us assume we wish to design a counter that will count $0, 1, 2, 3, 0, \ldots$. The first step is to draw the state diagram [Fig. 7-12(a)]. Next, we shall draw a truth table for two FF's, A and B. Below the line, we must supply two control signals: G_A, for controlling whether A flips, and G_B, for controlling whether B flips. Assume we are now in state 0. Then we must next go to state 1. However, to get there, A must not flip and B must flip. Therefore, we must supply a G_A signal so that

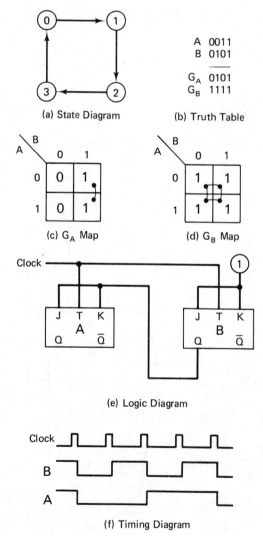

(a) State Diagram

$$\begin{array}{ll} A & 0011 \\ B & 0101 \\ \hline G_A & 0101 \\ G_B & 1111 \end{array}$$

(b) Truth Table

(c) G_A Map

(d) G_B Map

(e) Logic Diagram

(f) Timing Diagram

FIG. 7-12 Type T Modulo Four Counter

A will not flip (a 0) and a G_B signal so that *B* will flip (a 1) when we receive the next clock pulse. Consequently, we shall define the output of G_A to be a 0 when we are in state 0 and the output of G_B to be a 1 when we are in state 0.

Upon receipt of the next clock pulse, we shall be in state 1. But how do we get from 1 to 2? Flip-flop *A* must change state, as must FF *B*. Therefore, when we are in state 1, G_A should output a 1 and G_B should output a 1. Continuing this process, G_A must be 0 and G_B a 1 in the 2 state and G_A a 1 and G_B a 1 in the 3 state. Each time we are defining what the *G* lead to the FF must be in order to take us to the next state.

We have now defined the truth table; all we have to do is read the map. However, by this time, you should be able to read simple expressions directly off the truth table without mapping. Doesn't G_A look like *B* in its patterns of 1's and 0's? Why map it if this is apparent (you may if you want to)? G_B is a 1 all the time. We now have our two equations:

$$G_A = B$$

$$G_B = 1 \quad \text{(wire it high)}$$

All we have to do is draw the logic diagram [Fig. 7-12(e)]. Note that *A* is the MSB; this is consistent throughout this book. Take time out to follow the timing diagram. It will greatly increase your understanding of this and the following design techniques. At first, *A* and *B* are both 0. But *B* has a 1 on its *JK* lead; therefore it will flip next time. *A*, however, has a low and will not flip. Therefore, state 01 is entered. Now, both *A* and *B* are programmed to flip, so *B* goes from 1 to 0 and *A* from 0 to 1, entering state 2. *B* is low again, so *A* will not flip, and, upon receipt of the next clock pulse, state 3 is entered. Now, with both *A* and *B* high, both will flip to the 0 state.

Down Counters

This procedure can be used for down counting also. All that has to be done is to analyze what state each FF must enter upon receipt of the next clock pulse. If the FF is to remain in its present state, *G* must be a 0; if it is to flip, *G* must be a 1.

Example 7-3: Design a four-state down counter using type *T* design procedures.

SOLUTION:

First draw the state diagram [Fig. 7-13(a)] and then the truth table. Starting out in state 0, we must go to state 3. This will require both G_A and G_B to be 1's so both FF's will flip. Next, from state 3 we must go to 2, requiring *A* to remain unchanged (G_A a 0) and *B* to flip (G_B a 1). This continues until the entire table is completed. Reading the equations from table,

$$G_A = \bar{B}$$

$$G_B = 1$$

The final logic and timing diagrams are shown in Fig. 7-13(c) and (d).

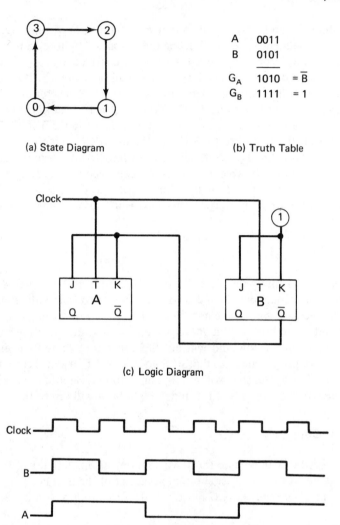

(a) State Diagram   (b) Truth Table

A	0011	
B	0101	
G_A	1010	$= \overline{B}$
G_B	1111	$= 1$

(c) Logic Diagram

(d) Timing Diagram

FIG. 7-13 Example 7-3, Type T Down Counter

Nonsequential Counting

The type T design can be used for any state counting. Assume we wanted to count 2, 7, 5, 6, 2, ... First, draw the state diagram [Fig. 7-14(a)] and then the truth table. You may find it helpful to draw state arrows at the top of the table. G_A and G_B can be read directly off the table; G_C was mapped. Armed with the logic equations, we can now draw the logic diagram. Note that, in order to use **NAND** logic, the **OR**

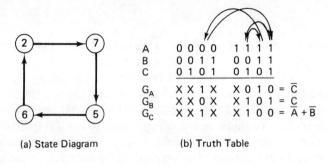

(a) State Diagram   (b) Truth Table

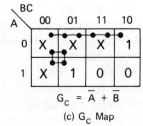

$G_C = \overline{A} + \overline{B}$

(c) G_C Map

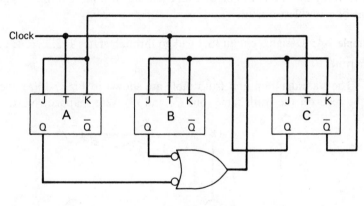

(d) Logic Diagram

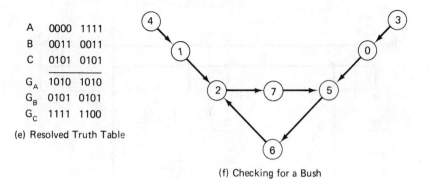

(e) Resolved Truth Table

(f) Checking for a Bush

FIG. 7-14 Nonsequential Type T Design

gate inputs were taken off A and B, rather than $\bar{A}$ and $\bar{B}$. By the time they are inverted by the gate inputs, they become $\bar{A}$ and $\bar{B}$, as the equation requires.

There is one more step we must take before considering the problem solved. What would happen if the circuit were turned on and the first state it entered was the 3 state (a don't care)? If G_A, G_B, and G_C were all 0, it would stay in that state forever. What we must do is to check each don't care state to see if the circuit will eventually go into a do care state, for if it gets there, it will go through the sequence we specified. There are several ways this can be done: Analyze the truth tables, analyze the maps, or analyze the logic circuit. The truth table is the simplest, however. Therefore, redraw the truth table, putting 1's or 0's in place of all don't cares in accordance with the logic equations. Then, draw the state diagram. Next, imagine the counter to be in state 0, and analyze what next state it will enter. With G_A a 1, A will flip to 1; with G_B a 0, B will stay a 0; with G_C a 1, C will flip to a 1. It will therefore enter state $101 = 5$. Draw this on the state diagram. Continue the process until all states have been analyzed. If all lead eventually to the desired states, the state diagram is said to be a *bush*. If one is not obtained, a different set of equations must be derived by setting one of the don't cares to a 1. Many times this will involve an equation with more terms. Figure 7-15 is an example of a bushless circuit.

Example 7-4: Design a circuit that will go through states 1, 3, 7, 6, 1, . . .

SOLUTION:

The state diagram and truth table are shown in Fig. 7-16(a) and (b). G_C can be read off the truth table, but G_A and G_B were mapped. Next, the truth

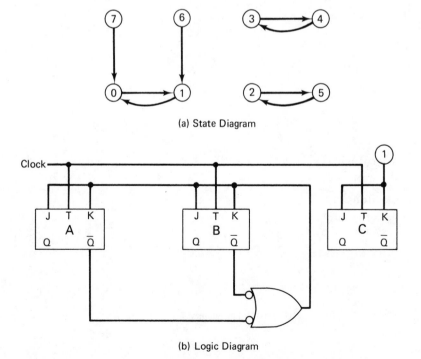

(a) State Diagram

(b) Logic Diagram

FIG. 7-15 Bushless Circuit

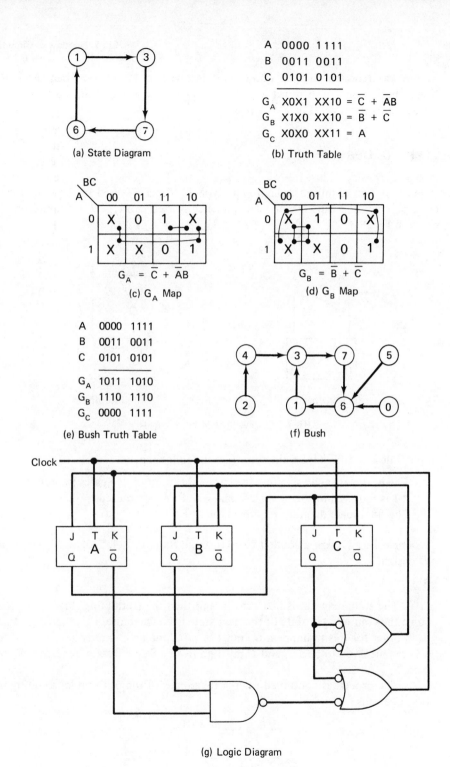

(a) State Diagram

A 0000 1111
B 0011 0011
C 0101 0101

G_A $\overline{X0X1 \ XX10}$ = $\overline{C}$ + $\overline{A}B$
G_B $X1X0 \ XX10$ = $\overline{B}$ + $\overline{C}$
G_C $X0X0 \ XX11$ = A

(b) Truth Table

$G_A = \overline{C} + \overline{A}B$

(c) G_A Map

$G_B = \overline{B} + \overline{C}$

(d) G_B Map

A   0000 1111
B   0011 0011
C   0101 0101

G_A 1011 1010
G_B 1110 1110
G_C 0000 1111

(e) Bush Truth Table

(f) Bush

(g) Logic Diagram

FIG. 7-16 Example 7-4

205

table was redrawn with all don't cares resolved to 1's or 0's. Then the bush was checked and drawn in Fig. 7-16(f). Since there was a bush, the final logic diagram was drawn [Fig. 7-16(g)].

7-5 TYPE *D* DESIGN

The type *D* design is quite similar to the type *T*. The FF's used in the design can either be purchased as type *D*'s or a *JK* can be converted to a type *D* by connecting it as shown in Fig. 7-17.

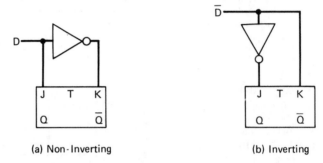

(a) Non-Inverting   (b) Inverting

FIG. 7-17 Converting *JK* FF to Type *D*

The type *D* design differs from the type *T* in the following:

A. The next state analysis using the truth table requires the gates to output a 1 if a 1 is required in the next state and a 0 if a 0 is required.
B. The final logic diagram requires type *D* FF's.

Example 7-5: Design a counter to go through states 0, 2, 4, 5, 0, . . . , using type *D* design.

SOLUTION:

The state diagram is first drawn, and then the truth table (Fig. 7-18). Assume the counter is in state 0; the next state is 2. Therefore, FF *A* must go to a 0; in order for this to happen, G_A must be a 0. Similarly, *B* must go from a 0 to a 1, requiring G_B to be a 1, and *C* must go from a 0 to a 0 state, requiring G_C to be a 0.

The process is continued for the remainder of the truth table, resulting in the equations

$$G_A = B + A\bar{C}$$
$$G_B = \bar{A}\bar{B}$$
$$G_C = A\bar{C}$$

(a) State Diagram

A   0000   1111
B   0011   0011
C   0101   0101

G_A 0X 1X   10XX = $B + A\bar{C}$
G_B 1X 0X   00XX = $\bar{A}\bar{B}$
G_C 0X 0X   10XX = $A\bar{C}$

(b) Truth Table

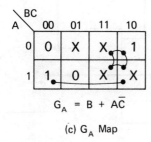

$G_A = B + A\bar{C}$

(c) G_A Map

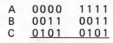

$G_B = \bar{A}\bar{B}$

(d) G_B Map

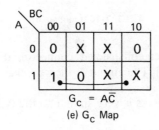

$G_C = A\bar{C}$

(e) G_C Map

A 0000   1111
B 0011   0011
C 0101   0101
_____
G_A 0011   1011
G_B 1100   0000
G_C 0000   1010

(f) Bush Truth Table

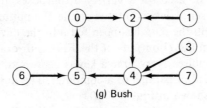

(g) Bush

FIG. 7-18 Example 7-5 Type *D* Design

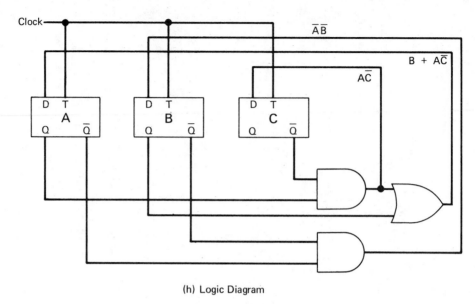

(h) Logic Diagram

FIG. 7-18 (continued)

The logic diagram is then constructed, connecting the G outputs to the D inputs of the FF [Fig. 7-18(h)]. The circuit is then checked for a bush.

As can be seen, the process is very similar to the type T design.

Shift Registers

A shift register is a device capable of shifting a binary word to the left or right. The most basic form is that of type D FF's connected together as shown in Fig. 7-19(a). If the switch were to close at the time shown, Q_A would shift in a 0 on the trailing edge of the clock pulse. This level would now appear at D_B and will be shifted into FF B at the trailing edge of the second clock pulse. Thus, the 0 is shifted to the right on each clock pulse. Notice that when the switch is opened the 1 is shifted to the right in a similar manner. JK FF's can also be used, and they react in an identical manner [Fig. 7-19(c)].

Shift registers can be used for a variety of purposes, one of which is serial to parallel conversion. The serial data is entered on the D input to FF A and then clocked into the register. When all the data has been fed into the device the clock is stopped and the data is read from the Q outputs of the FF's. For example, a serial keyboard generates 11 serial time pulses every time a key is depressed. These can be fed into a shift register and then read in parallel to a computer. The logic symbol for a serial to parallel shift register is shown in Fig. 7-20(a).

Data must also be converted from parallel to serial. Some printers require serial

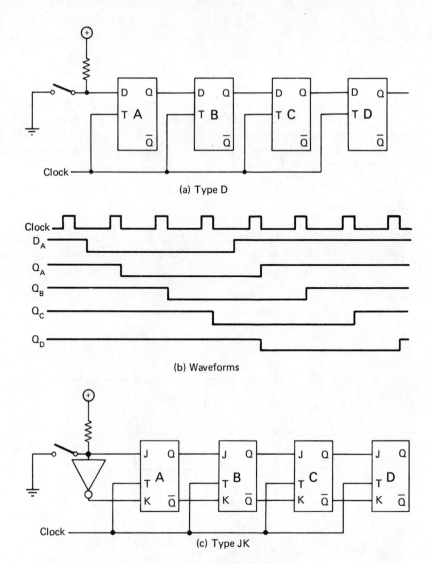

(a) Type D

(b) Waveforms

(c) Type JK

FIG. 7-19 Shift Registers

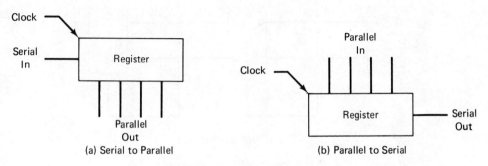

(a) Serial to Parallel

(b) Parallel to Serial

FIG. 7-20 Shift Register Logic Symbols

data, so the computer can store it in parallel in a shift register and then clock it out serially [Fig. 7-20(b)] from the Q output of the last stage. There are two methods of entering the parallel data into the register. One is to jam set (that is a correct digital term) or clear each of the registers as shown in Fig. 7-21(a), whenever the ENTER signal goes high. The second method is to clock it in after the SERIAL signal goes low in Fig. 7-21(b). In this diagram, if the SERIAL signal is low, the data is loaded in parallel; if it is high, the data in the register is shifted to the right.

Ring Counters

There are a number of computer systems that use a ring counter for their basic timing. This device is a shift register that shifts a single 1 from its input to its output.

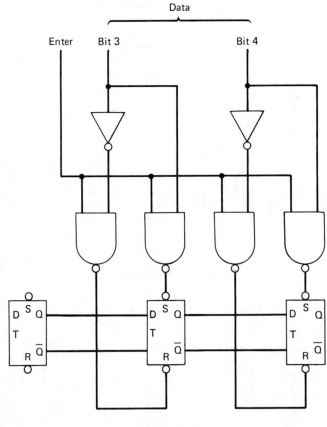

(a) Jam Set/Reset

FIG. 7-21 Parallel Loading a Shift Register

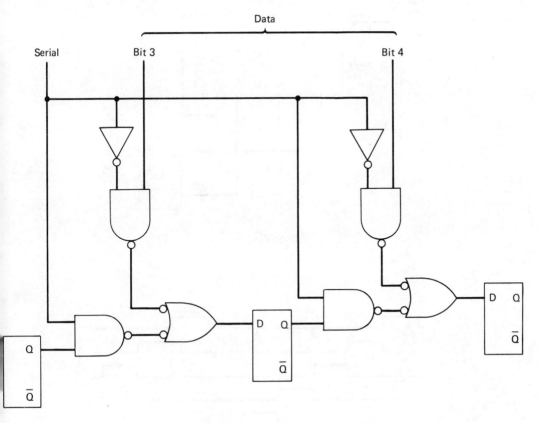

(b) Clocked Data Entry

FIG. 7-21 (continued)

The "ring" part comes from the fact that the output can be fed back into the input to form a ring. If only one 1 were circulating, the waveform would be as shown in Fig. 7-22(b). This is a very useful waveform, for any number of FF's could be connected as FF *L* is to produce any gating signal necessary. Note that the entire diagram has a minimum of gating, reducing propagation delays, and is purely synchronous. This specific circuit has the disadvantage of accumulating errors. If two 1's get into the ring, they would circulate forever. However, this can be overcome by (a) resetting the counter after each pass or (b) **OR**ing all the outputs such that if there is a 1 in the ring, a 0 will be fed into the input.

One of the disadvantages of the ring counter is that it uses one FF per state; to have 16 unique states requires 16 FF's. To overcome this, a complementing ring counter may be used (Fig. 7-23). Assume it starts out with no 1's in the counter; the waveforms would be as shown. Note that with four FF's, 8 unique states are obtained. This type of counter will obtain two times as many states as FF's.

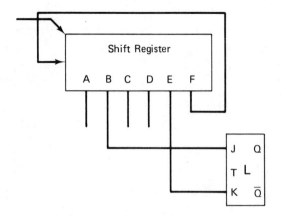

(a) Logic Diagram

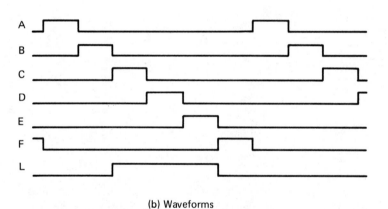

(b) Waveforms

FIG. 7-22 Ring Counter

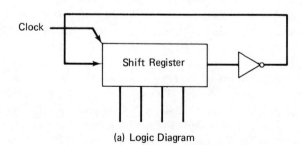

(a) Logic Diagram

FIG. 7-23 Complementing Ring Counter

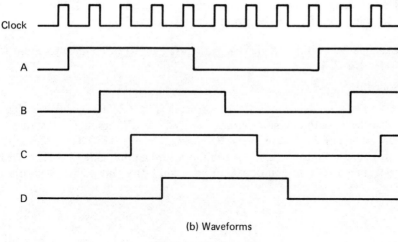

(b) Waveforms

FIG. 7-23 (continued)

7-6 TYPE *JK* DESIGN

The *JK* FF is a combination of a type *T* and a type *D*. Therefore, it is possible to take advantage of this duality and reduce the amount of gating logic. For example, assume that a *JK* FF is in the 0 state and that it must go to the 1 state. There are two ways this may be done:

    A. Apply a logical 1 to the *J* lead and a logical 0 to the *K* lead. This will shift in a 1 in the type *D* mode.

    B. Apply a 1 to both the *J* and *K* leads and flip it to the 1 state using the type *T* mode.

But what have we observed? To change the *JK* FF from the 0 to the 1 state a 1 must be applied to the *J* lead and either a 0 or a 1 to the *K* lead. Therefore, the *J* must be a 1 and the *K* a don't care. This same reasoning applies to any state transition, all of which are summarized in Table 7-1. For example, to go from the 1 state to the 1 state, the *J* lead may be either a 1 or 0 and the *K* lead a 0.

    Using this capability of the *JK* FF, at least one-half the entries on a truth table will be don't cares, greatly simplifying the circuit.

TABLE 7-1. *JK* Flip-flop Input Requirements

	State Transitions			
	0 → 0	0 → 1	1 → 0	1 → 1
J	0	1	X	X
K	X	X	1	0

The *JK* design procedure requires one gate per each *J* or *K* lead (two per FF). However, some of these may be straight wires.

Example 7-6: Using *JK* design procedures, design a counter to count through states 1, 2, 3, 1, ...

SOLUTION:

The state diagram is first constructed [Fig. 7-24(a)]. The truth table is then constructed, with one gate designed to feed each *JK* input. The 0 state is a don't care. When in the 1 state, the *A* FF must go from a 0 to a 1; according to Table 7-1, the *J* lead must be a 1 (J_A) and the *K* lead a don't care (K_A). The *B* FF must go from a 1 to a 0, requiring J_B to be a don't care and K_B a 1, according to the

(a) State Diagram

A   0011
B   <u>0101</u>

J_A X 1XX = 1
K_A XX 01 = B
J_B XX 1X = 1
K_B X 1X0 = $\overline{A}$

(b) Truth Table

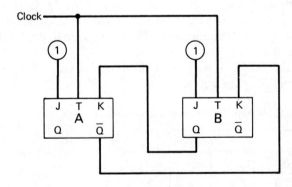

(c) Logic Diagram   **FIG. 7-24** Example 7-6, *JK* Design

table. In state 2, the *A* FF must go from a 1 to a 1; therefore, J_A must be an *X* and K_A a 0. *B* must go from a 0 to a 1; J_B must, therefore, be a 1 and K_B an *X*. Similarly, in state 3, J_A must be *X*, K_A a 1, J_B an *X*, and K_B a 0.

Each output is then converted to an equation from the truth table (use a map, if necessary; however, this one is very easy). The logic diagram is constructed from these equations. Finally, we must see that the 0 state leads to a bush. This can be done directly from the circuit, for if both *A* and *B* are 0, both *J* leads will still be 1, causing both to flip, upon receipt of the next clock pulse.

Example 7-7: Design a *JK* counter for states 0, 8, 9, 5, 10, 12, 14, 15, 0, ...

SOLUTION:

First, the state diagram was drawn, and then the truth table (Fig. 7-25). Each state was analyzed for the state into which it must enter. The equations were then derived from the truth table and the maps and the logic diagram drawn. Finally, each don't care state was analyzed for a bush.

Example 7-8: Design a pulse train generator for the following waveform:

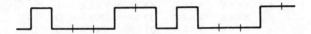

SOLUTION:

The pulse train is first copied down and then assigned states [Fig. 7-26(a) and (b)]. Note that we are taking the waveform off FF *C* and are arbitrarily assigning *A* and *B* to obtain unique states. This results in the seven unique states shown in Fig. 7-26(c). Using *JK* design, the final logic diagram is shown in Fig. 7-26(e).

Controlling the Counter

In some applications a counter must count through more than one set of states as determined by control lines. These control lines are then added to the truth table, and the logic is designed using the control line terms as inputs to the combinational logic.

Example 7-9: Design a *JK* counter that must go through states 0, 3, 6, 0, . . . if a control line is high and 0, 2, 4, 6, 0, . . . if the control line is low.

SOLUTION:

The combinational logic will have four inputs: FF's *A*, *B*, and *C* and control line *X*. These are entered on the truth table [Fig. 7-27(a)] and the equations solved. The logic diagram is then constructed and a check made for a bush. In this case, it must have bushes for each mode of operation.

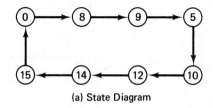

(a) State Diagram

A	0000	0000	1111	1111
B	0000	1111	0000	1111
C	0011	0011	0011	0011
D	0101	0101	0101	0101

J_A	1XXX	X1XX	XXXX	XXXX	= 1
K_A	XXXX	XXXX	010X	0X01	= D
J_B	0XXX	XXXX	011X	XXXX	= C + D
K_B	XXXX	X1XX	XXXX	0X01	= D
J_C	0XXX	X1XX	00XX	1XXX	= B
K_C	XXXX	XXXX	XX1X	XX01	= $\overline{B}$ + D
J_D	0XXX	XXXX	1X0X	0X1X	= $A\overline{B}\overline{C}$ + BC
K_D	XXXX	X1XX	X0XX	XXX1	= B

(b) Truth Table

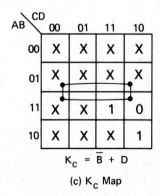

$K_C = \overline{B} + D$

(c) K_C Map

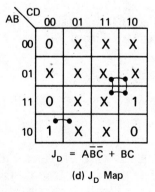

$J_D = A\overline{B}\overline{C} + BC$

(d) J_D Map

FIG. 7-25 Example 7-7, *JK* Design

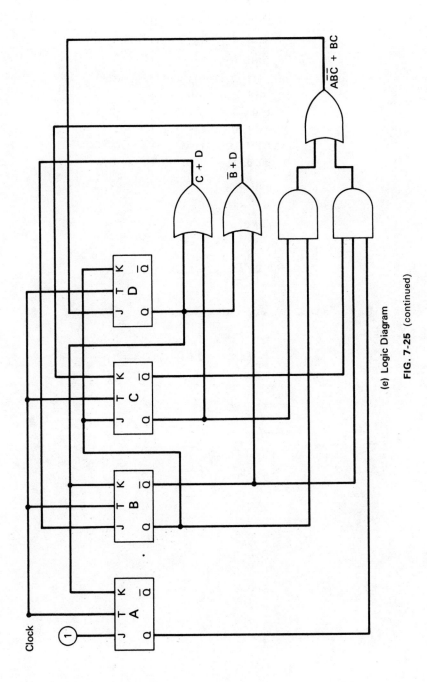

(e) Logic Diagram

FIG. 7-25 (continued)

A 0000 0000 1111 1111
B 0000 1111 0000 1111
C 0011 0011 0011 0011
D 0101 0101 0101 0101

J_A 1111 1111 1111 1111
K_A 0101 0101 0101 0101
J_B 0111 0111 0111 0111
K_B 0101 0101 0101 0101
J_C 0000 1111 0000 1111
K_C 1111 0101 1111 0101
J_D 0000 0011 1100 0011
K_D 0000 1111 0000 1111

(f) Bush Truth Table

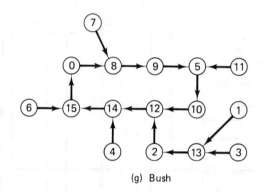

(g) Bush

FIG. 7-25 (continued)

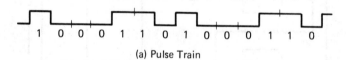

1 0 0 0 1 1 0 1 0 0 0 1 1 0

(a) Pulse Train

A	B	C	State
0	0	1	1
0	0	0	0
0	1	0	2
1	0	0	4
0	1	1	3
1	1	1	7
1	1	0	6

(b) State Assignment

FIG. 7-26 Example 7-8

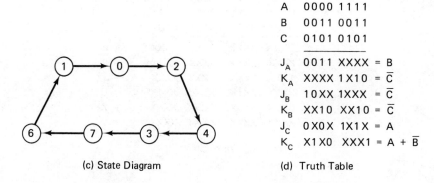

A	0000 1111
B	0011 0011
C	0101 0101

J_A	0011 XXXX	= B
K_A	XXXX 1X10	= $\overline{C}$
J_B	10XX 1XXX	= $\overline{C}$
K_B	XX10 XX10	= $\overline{C}$
J_C	0X0X 1X1X	= A
K_C	X1X0 XXX1	= A + $\overline{B}$

(c) State Diagram   (d) Truth Table

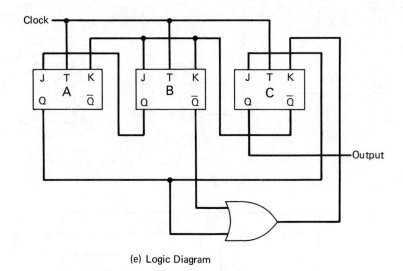

(e) Logic Diagram

FIG. 7-26 Continued

Cycle Counters

A cycle counter is a counter that outputs a stated number of counts and then stops. There are two basic circuits that can be used to control the counter.

The first, shown in Fig. 7-28, requires a ripple counter. The counter is started by providing a short negative pulse on the START line. This sets the CONTROL FF, which gates clock pulses into the counter. When the counter reaches its full count, it is reset by the feedback pulse; this same pulse resets the CONTROL FF, ending the sequence. The disadvantage of this type of circuit is that if the START line is operated at precisely the right moment, the CONTROL FF could turn on at a time to admit only the last edge of a clock pulse through the CLOCK gate. This could provide unpredictable results.

X	0000	0000	1111	1111	
A	0000	1111	0000	1111	
B	0011	0011	0011	0011	
C	0101	0101	0101	0101	
J_A	0X1X	XXXX	0XX1	XXXX	= B
K_A	XXXX	0X1X	XXXX	XX1X	= B
J_B	1XXX	1XXX	1XXX	XXXX	= 1
K_B	XX1X	XX1X	XXX0	XX1X	= $\overline{C}$
J_C	0X0X	0X0X	1XXX	XX0X	= $X\overline{B}$
K_C	XXXX	XXXX	XXX1	XXXX	= $\overline{A}$

(a) Truth Table

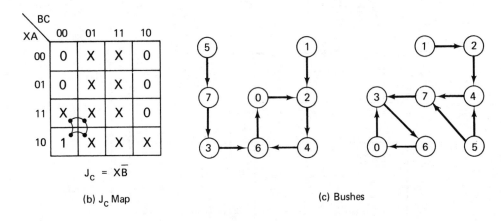

$J_C = X\overline{B}$

(b) J_C Map

(c) Bushes

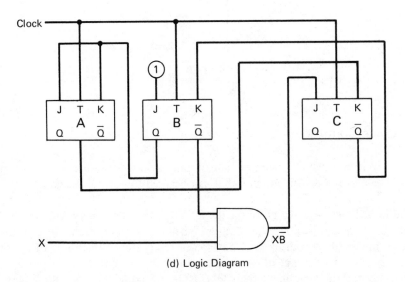

$X\overline{B}$

(d) Logic Diagram

FIG. 7-27 Example 7-9, Mode Control

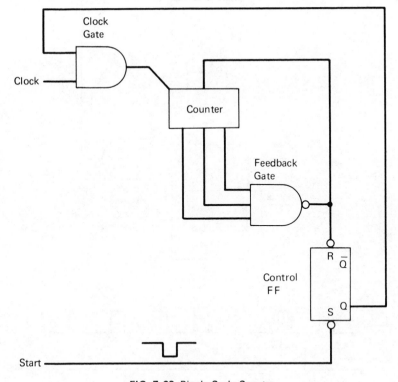

FIG. 7-28 Ripple Cycle Counter

A fully synchronous approach is shown in Fig. 7-29. In this circuit, the START line is energized and then clocked synchronously into the CONTROL FF. This opens the CLOCK gate, permitting clock pulses to be fed to the counter. Note that the control gate will open after the CONTROL FF has set, thus preventing any skinny pulses from being fed to the counter. The counter continues to count until it reaches its last state, which is decoded by the LAST STATE DECODER. With this gate high, the *K* lead of the CONTROL FF will clock a 0 into the FF on the trailing edge of the next clock pulse, turning off the CLOCK gate.

Example 7-10: Design a synchronous cycle counter that will go through states 0, 1, and 2 each time a start pulse is issued.

SOLUTION:

First, the counter must be designed (Fig. 7-30) using *JK* design procedures. The CONTROL FF and CLOCK gate were then added. Note that the last state is a 2, a binary 10, and that the MSB, FF *A*, is only 1 during this last state. It can therefore be used to gate the CONTROL FF off.

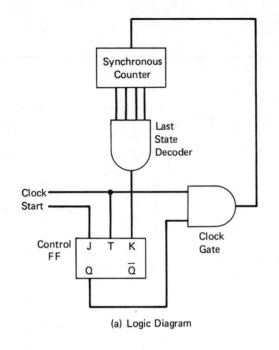

(a) Logic Diagram

(b) Timing Diagram

FIG. 7-29 Synchronous Cycle Counter

A	0011
B	<u>0101</u>
J_A	01XX $= B$
K_A	XX1X $= 1$
J_B	1X0X $= \overline{B}$
K_B	X1XX $= 1$

(a) State Diagram   (b) Truth Table

FIG. 7-30 Example 7-10

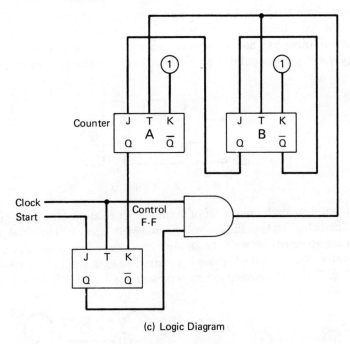

(c) Logic Diagram

FIG. 7-30 Continued

7-7 ASYNCHRONOUS SEQUENTIAL CIRCUITS

The sequential circuits discussed thus far have all used clocks and were, therefore, synchronous circuits. These circuits have the advantage of being free from any logic race or propagation delay problems. However, they all have the disadvantage of being required to wait for receipt of the next clock pulse before proceeding. Asynchronous circuits are therefore faster—messier, for they are subject to logic race problems—but faster, nevertheless.

The general model for asynchronous circuits is shown in Fig. 7-31, where the inputs are A_1 and A_2, the outputs are Z_1 and Z_2, and the "remembering the previous

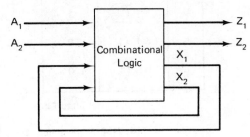

FIG. 7-31 Asynchronous Circuit Model

state" feedback lines are X_1 and X_2. It should be emphasized that all this logic is combinational. The only difference is that some of the outputs are fed back to the inputs.

Assume an input line has the waveform shown below and it is required that the output display every other pulse. You might call this a pulse gulping circuit.

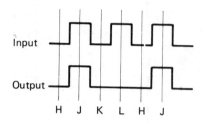

To solve the problem, each unique set of input/output relationships is assigned a state as shown. The actual state numbers will be assigned later. It doesn't matter if too many states are assigned, for we have an algorithm for reducing them to a minimum number of states. Next, a state diagram is constructed with input and output information on it (Fig. 7-32). The circled letters represent the states as before. However, an

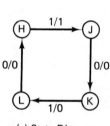

(a) State Diagram,
First Step

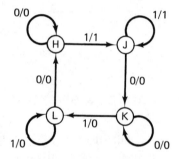

(b) Completed State Diagram

State	A Input		Z Output	$X_1 X_2$ State Assignments
	0	1		
H	(H)	J	0	01
J	K	(J)	1	11
K	(K)	L	0	10
L	H	(L)	0	00

(c) State Table

FIG. 7-32 Asynchronous Logic States

identifier is included, indicating the condition of the inputs and the outputs necessary to take it to the next state. In this identifier, 1/1, the left binary word (1) represents the condition of the input and the right word (1) the condition of the output when going from state H to J. Notice that from state J a 0 into the circuit will send the circuit to state K, outputting a 0. From state K, a 1 in will cause the circuit to go to state L, outputting a 0, and from L a 0 in will send it to state H with a 0 output.

Although the next procedure may seem unnecessary, it is essential in order to construct the truth table. Let us now assume we are in state H and a 0 is inputted. Where will the circuit go? The answer is, of course, state H, so a loop is drawn back to state H (Fig. 7-32). Then state J is analyzed. What if a 1 is inputted? It will stay in J and output a 1, so a second loop is drawn back to J. The procedure is repeated for all the states.

Next, the state table is drawn. In this table, each state is listed along the left-hand side and every possible input along the top. Assume the circuit is in state H and it receives a 0 input. According to the state diagram, it will stay in that state—until eternity, if necessary. Therefore, this is called a stable state, and the state letter in which it will stay is entered and circled. Now, assume a 1 is received; according to the state diagram, it will go to state J, so enter a J under the 1's input column in state H. However, it will now go to state J, instead of staying in its present state, so do not circle it. Then move to the 1-input column of the J state. A 1 is now being received; where will it go? It will, of course, stay in the J state, so this is a stable state; circle it. When a 0 is received while in the J state, the circuit will again change states by moving to state K, so enter a K. This process is continued until all states have been analyzed. The output is also entered in the output column. In this problem, the only time an output is required is during state J. It must be 0 during all other states.

The next step is to assign binary numbers to each state. Try to assign them so that only one variable will ever change when the circuit goes from a particular state to the next. These are shown in the assignment column. Note that a Gray code satisfies this requirement and that there need be no relationship between the initial lettered states assigned and these binary assignments.

The block diagram for the final circuit is shown in Fig. 7-33(a). There will be three inputs, A, X_1, and X_2. Therefore, a truth table is constructed for the three input and output variables. Do not worry about the fact that an input may also be an output. Remember that as long as we can define the inputs we can define any output. Let us now consider when $X_1X_2A = 000$; that is, we are in state L and have a 0 on A. According to Fig. 7-32(c), we want to go to state H, a 01 in terms of X_1X_2. Therefore, enter 0 for X_1 output and 1 for X_2 output. Next, examine $X_1X_2A = 001$. This means we are in state L and receive a 1 input. In this condition, the circuit must stay in L, so enter 00 into X_1X_2 output terms. $X_1X_2A = 010$ is next; this represents state H with a 0 input. According to the state diagram, the circuit must remain in H, so enter a 01 for the X_1X_2 outputs. This process continues until the truth table has been filled out. The final entry on the truth table is that of the output, Z. This must be 1 anytime state J is encountered, so enter 1's under any $X_1X_2 = 11$ input terms.

Having drawn the truth table, it is now a simple matter to come up with the out-

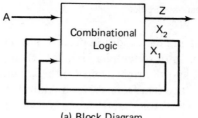

(a) Block Diagram

X_1	0000	1111
X_2	0011	0011
A	0101	0101
X_1	0001	1011
X_2	1011	0001
Z	0000	0011

(b) Truth Table

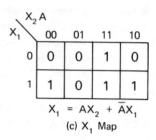

$X_1 = AX_2 + \overline{A}X_1$

(c) X_1 Map

$X_1 \backslash X_2 A$	00	01	11	10
0	1	0	1	1
1	0	0	1	0

$X_2 = AX_2 + \overline{A}X_1$

(d) X_2 Map

$X_1 \backslash X_2 A$	00	01	11	10
0	0	0	0	0
1	0	0	1	1

$Z = X_1 X_2$

(e) Z Map

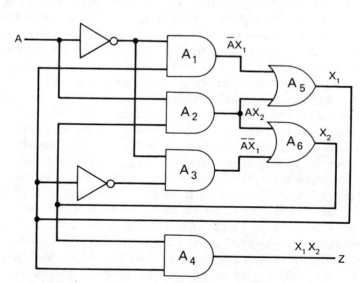

(f) Logic Diagram

FIG. 7-33 Asynchronous Logic, Mapping

226

put equations:

$$X_1 = AX_2 + \bar{A}X_1$$
$$X_2 = AX_2 + \bar{A}\bar{X}_1$$
$$Z = X_1X_2$$

The final step is to draw the logic diagram [Fig. 7-33(f)].

Having developed the circuit, let us now analyze it. Assume X_1, X_2, and A are low. Gate A_1 cannot AND, for X_1 is low; gate A_2 cannot AND for A is low. However, A_3 can AND, causing X_2 to go high. This has no further effect and was entered into Table 7-2. Now, assume A goes high. This will turn on A_2, causing both X_1 and X_2 to

TABLE 7-2. Asynchronous Circuit State Analysis

X_1X_2	A 0 1	Z
00	00	0
01	01	0
11	11	1
10	10	0

go high, putting a 1 on Z. Now assume A goes low again. Gates A_3 and A_2 will not AND, but A_1 will, putting the system in the 10 state and removing the 1 from Z. Next, A_1 will go high, cutting off A_1, A_2, and A_3 and sending the circuit into the 00 state. When A goes low again, the circuit will return to the 01 state.

The entire design procedure can be summarized:

A. Draw waveforms.
B. Draw the state diagram.
C. Draw the state table.
D. Simplify the state table. This was not necessary in the previous problem.
E. Assign binary states.
F. Draw the truth table.
G. Map.
H. Implement the logic equations.

Example 7-11: Design a type T flip-flop from logic gates.

SOLUTION:

A block diagram is shown in Fig. 7-34(a). First, the waveforms are drawn, trying to take into account all possible conditions. Then they are assigned lettered states [Fig. 7-34(b)]. The state diagram is drawn next [Fig. 7-34(c)]. Note that 1's must be outputted in both states C and D. The truth table is then drawn, the map evaluated, and the final circuit designed [Fig. 7-34(i)].

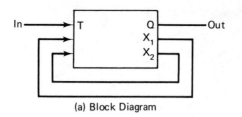

(a) Block Diagram

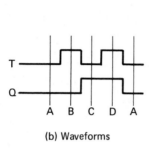

(b) Waveforms

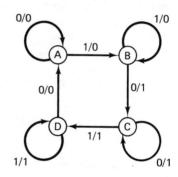

(c) State Diagram

State	T 0	T 1	Q	State Assignment
A	(A)	B	0	00
B	C	(B)	0	01
C	(C)	D	1	11
D	A	(D)	1	10

(d) State Table

X_1 0000 1111
X_2 0011 0011
T   <u>0101</u> <u>0101</u>
X_1 0010 0111
X_2 0111 0010
Q   0000 1111

(e) Truth Table

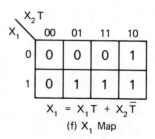

$X_1 = X_1 T + X_2 \overline{T}$

(f) X_1 Map

FIG. 7-34 Example 7-11, Type T Flip-flop

228

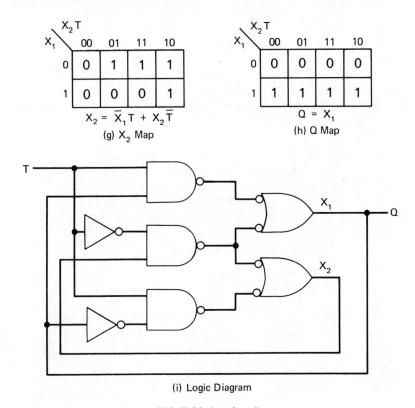

(i) Logic Diagram

FIG. 7-34 (continued)

State Reduction

In the previous section, we were fortunate enough to choose an optimal set of states. However, in many problems, the optimal set is not apparent. Let us reexamine Example 7-11 and purposefully choose a nonoptimal set so the state reduction process can be demonstrated. The waveforms will be redrawn and new lettered states assigned [Fig. 7-35(a)]. Next, the state diagram must be drawn [Fig. 7-35(b)], including all states.

To reduce the state diagram, a table is constructed, called an *implication table* [Fig. 7-35(d)]. It allows us to compare every lettered state with every other lettered state for possible merging. To merge, the output states must be identical. First, examine *A* and *B* in the state table. If *T* is a 0, they can be merged only if *A* equals *C*. If *T* is a 1, they can be merged only if *B* equals *B* (which it does). Therefore, we shall enter an *AC* into the *AB* square [Fig. 7-35(e)]. Next, examine *A* and *C*. They cannot be merged because their outputs differ, so cross out the *AC* square. *A* and *D* produce the same result. *A* and *E* can be merged if *A* and *E* are identical (if we merged them, this would be a true statement) and if *B* and *F* are identical. Therefore, enter *BF*. *A* and *F* will merge if *A* and *G* merge and if *B* and *F* merge. This should continue until all terms have been compared with all other terms. Note that *C* and *G* are already identical, so enter a check in that square.

Now, perform the second pass. *A* and *B* can be merged only if *A* and *C* can be merged. But *A* and *C* are incompatible (see the *X* in the *AC* square?); therefore, *A* and *B* cannot be merged, so cross out the *AC* square [Fig. 7-35(f)]. *A* and *E* require *B* and *F* to be compatible, and they are still in the running, so leave that square as it is.

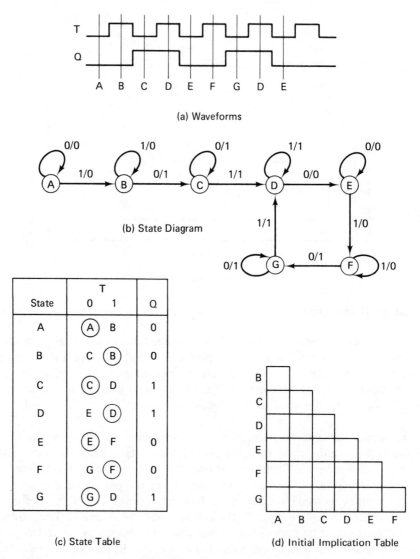

(a) Waveforms

(b) State Diagram

(c) State Table

(d) Initial Implication Table

FIG. 7-35 State Reduction

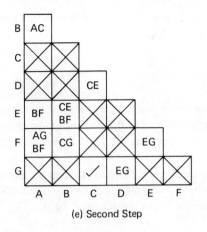

(e) Second Step

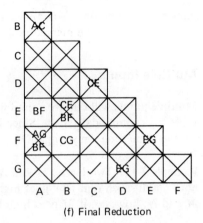

(f) Final Reduction

F

E

D

C   (C,G)

B   (C,G) (B,F)

A   (A,E) (C,G) (B,F)

(g) Collecting Compatibility Classes

States	T		Q
	0	1	
A	(A)	B	0
B	C	(B)	0
C	(C)	D	1
D	A	(D)	1

(h) Reduced State Table

FIG. 7-35 (continued)

However, *A* and *F* cannot merge because *A* and *G* cannot, so cross out the *AF* square. This process continues until all squares that cannot go together have been crossed out.

Next, start with the *F* column and collect *go-togethers* [Fig. 7-35(g)]. There are none in the *F* column, nor the *E*, nor *D*. However, *C* has *C* and *G*. Accumulate these compatibility classes until the *A* column is finished. Since no member of one class appears twice, the states can be merged as follows:

C with *G*

B with *F*

A with *E*

The next step is to actually perform the merger on the state table, eliminating redundant rows. Every time a G appears, replace it with a C; F's with B's and E's with A's. This results in the merged table [Fig. 7-35(h)].

Multiple Inputs

Multiple inputs will require a modification of the state diagram, state table, and truth table to accommodate them. Since it is impossible for two inputs to change at precisely the same time, only single Gray-code-type changes need be considered.

Example 7-12: Two levels, M and N, enter a logic circuit. Design an asynchronous circuit such that if N goes high before M, the output will be the **AND** of M and N; however, if M goes high before N the output must be 0.

SOLUTION:

Analysis of the possible input states produces some interesting conclusions (Fig. 7-36). Note that each possible case has to be considered: when M occurs before N, after N, and during N. Therefore, lettered states are assigned as shown. The state diagram must be drawn next. The easiest way to complete it is to take the information directly off the pulse analysis waveforms and then go back over the state diagram and consider what happens as each bit changes. For example, state C is entered with a 11/1 (the 11 means the MN inputs are 1 and 1, respectively; the right-hand 1 means the output is 1). There are only two possible changes from the 11 state, 10 and 01; therefore, both must be considered.

Next, the state table must be constructed. Note that there are now four input columns to accommodate all four states of MN. Consider the A state (00). There are only two possible states into which it could go: 01 or 10. Therefore, put a dash in the 11 column and the proper letters in the other columns. This is continued until the table is completed.

The states must now be reduced to the minimum necessary to do the job. In making entries on the implication table, a letter can always be merged with a dash. The implication table is shown in Fig. 7-36(d).

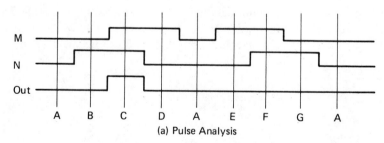

(a) Pulse Analysis

FIG. 7-36 Example 7-12, Multiple Inputs

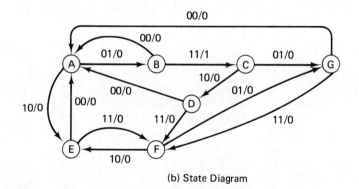

(b) State Diagram

State	MN				Out
	00	01	10	11	
A	(A)	B	E	–	0
B	A	(B)	–	C	0
C	–	G	D	(C)	1
D	A	–	(D)	F	0
E	A	–	(E)	F	0
F	–	G	E	(F)	0
G	A	(G)	–	F	0

(c) State Table

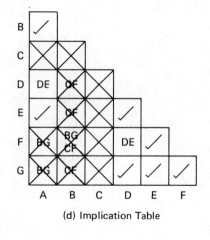

(d) Implication Table

F (FG)

E (EFG)

D (DEFG)

C (DEFG)

B (DEFG)

A (DEFG) (ADE) (AB)

(e) Compatibility Classes

FIG. 7-36 (continued)

State	MN Inputs 00	01	10	11	Output	Binary State $X_1 X_2$
A	Ⓐ	Ⓐ	D	C	0	00
C	—	D	D	Ⓒ	1	01
D	A	Ⓓ	Ⓓ	Ⓓ	0	11

(f) Reduced State Table

State	MN Inputs 00	01	10	11	Output	Binary State $X_1 X_2$
A	Ⓐ	Ⓐ	H	C	0	00
C	A	D	D	Ⓒ	1	01
D	C	Ⓓ	Ⓓ	Ⓓ	0	11
H	A	—	D	—	0	10

(g) Final State Table Eliminating Races

```
X₁   0000   0000   1111   1111
X₂   0000   1111   0000   1111
M    0011   0011   0011   0011
N    0101   0101   0101   0101

X₁   0010   0110   0X1X   0111
X₂   0001   0111   0X1X   1111
Z    0000   1111   0000   0000  = X̄₁X₂
```

X_1 0000 0000 1111 1111
X_2 0000 1111 0000 1111
M 0011 0011 0011 0011
N 0101 0101 0101 0101

X_1 0010 0110 0X1X 0111
X_2 0001 0111 0X1X 1111
Z 0000 1111 0000 0000 $= \bar{X}_1 X_2$

(h) Truth Table

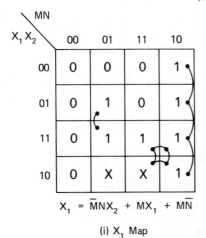

$X_1 = \overline{M}NX_2 + MX_1 + M\overline{N}$

(i) X_1 Map

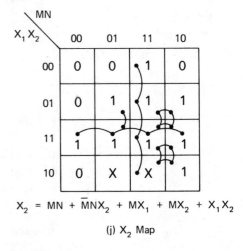

$X_2 = MN + \overline{M}NX_2 + MX_1 + MX_2 + X_1X_2$

(j) X_2 Map

FIG. 7-36 (continued)

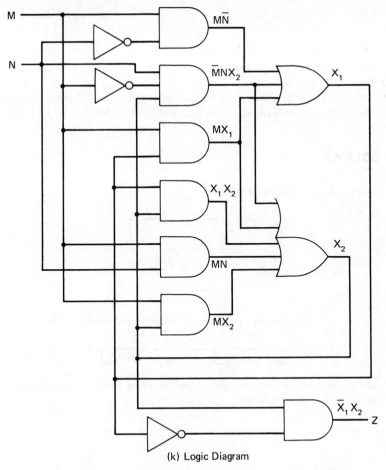

(k) Logic Diagram

FIG. 7-36 (continued)

7-8 SUMMARY

Sequential logic requires that inputs be supplied both in the proper combination and in the proper time sequence. There are four logic FF's used for this purpose:

    A. Reset-set (RS)
    B. Toggle (T)
    C. Delay (D)
    D. Type JK

These provide the basis for four counter design techniques:

    A. Ripple counters
    B. Type T design
    C. Type D design
    D. Type JK design

The most widely used are the ripple counter, because of its ease of design and simple circuits, and the *JK* design, because of its relative ease of design, wide availability, and purely synchronous output.

Shift registers can be used for serial/parallel operations. Asynchronous logic is inherently faster than synchronous logic but is subject to logic race problems.

7-9 PROBLEMS

7-1. Draw the waveform of the Q output:

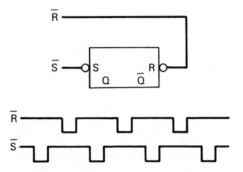

7-2. Draw the waveform of the Q output of the following circuit in which the FF flips on a negative to positive transition:

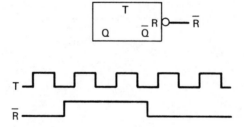

7-3. Draw the Q output waveform of the following circuit in which the FF shifts on a negative to positive transition:

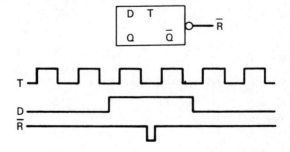

7-4. Draw the waveform of the Q output if the FF activates on a positive to negative transition:

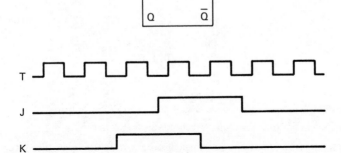

7-5. Design a divide-by-6 ripple up counter using FF's that toggle on a positive to negative transition. The outputs must be taken off the Q leads of the FF's.

7-6. Design a divide-by-14 ripple up counter using counters that toggle on negative to positive transitions. Take the output off the Q leads.

7-7. Design a divide-by-6 ripple down counter that counts down from state 7. Use FF's that toggle on positive to negative transitions, and take the outputs off the Q leads.

7-8. Design a divide-by-10 ripple down counter that counts down from 15. Take the outputs off the Q lead.

7-9. Design a divide-by-6 type T up counter that omits states 0 and 1.

7-10. Design a type T counter that goes through states 0, 3, 5, 6, 0, ...

7-11. Design a type D counter that goes through states 0, 1, 2, 4, 0, ...

7-12. Design a JK counter that goes through states 3, 4, 6, 7, 3, ...

7-13. Design a JK counter that goes through states 1, 2, 4, 5, 7, 8, 10, 11, 1, ...

7-14. Design a JK counter that goes through states 7, 4, 8, 3, 9, 2, 10, 1, 11, 0, 7, ...

7-15. Design a JK counter that goes through states 1, 2, 3, 6, 7, 8, 11, 12, 13, 1, ...

7-16. Design a JK counter that goes through states 25, 20, 15, 10, 5, 0, 25, ...

7-17. Design a JK counter to generate the following pulse train:

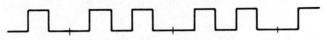

7-18. Design a JK counter to generate the following pulse train:

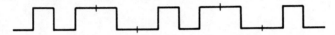

7-19. A three-stage JK counter must be designed that will count in two different modes, depending on the logic level of a control line. If the control line is high, the counter must count 0, 2, 4, 6, 0, ... If the control line is low, it must up-count through all eight states.

7-20. A two-stage *JK* counter has two input control lines, *M* and *N*. It must count in the following manner:

MN

00	0, 1, 2, 3, 0, . . .
01	0, 1, 0, 1, . . .
10	2, 0, 2, 0, . . .
11	1, 2, 1, 2, . . .

Design the circuit.

7-21. Design a cycle counter that will output states 0 through 5.

7-22. Design an asynchronous circuit that will output only the first pulse received. Any further pulses will be ignored.

7-23. Design an asynchronous circuit that will output only the second pulse received and will ignore any other pulses.

7-24. An asynchronous circuit has two inputs, *M* and *N*. Design a circuit that will output a 1 upon the first high-to-low or low-to-high transition of the *N* line after the *M* line has gone high.

7-25. Design an asynchronous circuit that will output pulses received on the *N* line only after a complete positive pulse has been received on the *M* line.

DIGITAL
INTEGRATED
CIRCUITS

8

8-1 MSI AND LSI

Although the basic elements of digital electronics are **AND** gates, **OR** gates, inverters, and flip-flops, MOS and TTL integrated circuit technology has provided many medium- and large-scale integrated circuits (MSI and LSI) that greatly reduce package count, saving both time and money (that long green stuff). However, they can be used only if the designer is familiar with the functions that are performed by these ICs (integrated circuits) and their characteristics. In this chapter we shall introduce these functions, describing and illustrating their use.

The TTL logic family is the most popular of those on the market today. The 7400 series TTL has become an industry standard. Manufacturers of CMOSs have also developed a pin-for-pin compatible family, calling it the 74C00 series, in which a TTL 7404 has the same pin assignments as a CMOS 74C04, for example.

8-2 LOGIC GATES

Several different types of gates are available to the designer, varying from simple **NAND** and **NOR** gates to Schmitt triggers and **AND/OR** functions. **AND, OR, NAND, NOR**, and exclusive **OR (XOR)** gates are all readily available.

TTL logic provides three types of gate outputs: totem pole, open collector, and tri-state. The totem pole outputs are used for general purposes and are the types illustrated in Fig. 8-1. To provide for wired **OR**ing, open collector outputs are available to which the designer must add his own resistor to V_{cc} (Fig. 8-2). For those people who cannot make up their minds, tri-state logic is available, providing a totem

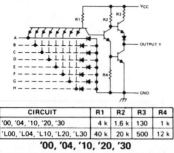

CIRCUIT	R1	R2	R3	R4
'00, '04, '10, '20, '30	4 k	1.6 k	130	1 k
'L00, 'L04, 'L10, 'L20, 'L30	40 k	20 k	500	12 k

'00, '04, '10, '20, '30
'L00, 'L04, 'L10, 'L20, 'L30, CIRCUITS
Input clamp diodes not on
SN54L'/SN74L' circuits.

FIG. 8-1 TTL Totem Pole Outputs, 7400 Series (Courtesy Texas Instruments, Incorporated)

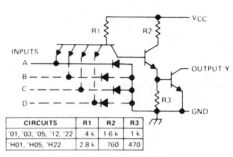

CIRCUITS	R1	R2	R3
'01, '03, '05, '12, '22	4 k	1.6 k	1 k
'H01, 'H05, 'H22	2.8 k	760	470

'01, '03, '05, '12, '22, 'H01, 'H05, 'H22 CIRCUITS

FIG. 8-2 TTL Open Collector Output, 7400 Series (Courtesy Texas Instruments, Incorporated)

pole output when the control lead is held low and an open circuit output when it is held high (Fig. 8-3). In this manner, the outputs of several devices can be wire-**OR**ed, as long as only one device has its control lead held low.

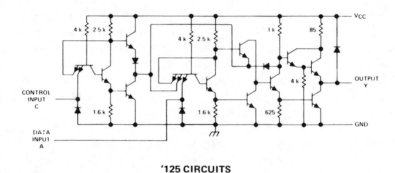

'125 CIRCUITS

FIG. 8-3 TTL Tri-State Totem Pole Output, 7400 Series (Courtesy Texas Instruments, Incorporated)

For those cases where extra drive is necessary, buffers can be used, providing up to 60-mA low-level output current in an open collector configuration; this is sufficient for driving certain low-current incandescent lamps and light-emitting diodes.

Since most combinational logic analysis results in a two-level function, **AND/OR**/invert gates are available [Fig. 8-4(top)], saving on package count. If the integrated circuit (IC) does not supply enough **AND** gates for feeding the **OR** gate, expandable gates are available to which an expander can be connected. Figure 8-4 shows such a

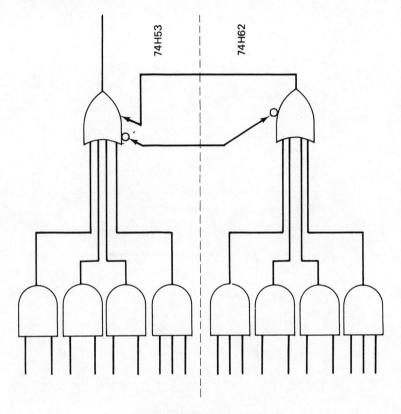

74H53

74H62

FIG. 8-4 Using Gate Expanders

circuit formed from a 74H62 expander and a 74H53 four-wide **AND/OR**/invert gate, creating an **OR** gate fed by eight **AND** gates. Expanders are also available for single gates, allowing the designer to assemble a gate with a great number of inputs.

The Schmitt Trigger

A Schmitt trigger gate is a very useful device for receiving digital signals from a line that may have noise. A typical Schmitt gate will recognize a logical 1 if the input voltage exceeds 1.7 V; however, once it has recognized this logical 1, the input must decrease to below 0.9 V to be recognized as a logical 0. This action is illustrated in Fig. 8-5(a) and (b). When the positive voltage first exceeds the 1.7-V threshold voltage the device outputs a logical 1. Note that even if the voltage decreases below the 1.7-V level, it will still output a logical 1. When the input first decreases to the 0.9-V level, the output goes low.

The symbol for the Schmitt trigger is shown in Fig. 8-5(c). There is a small figure

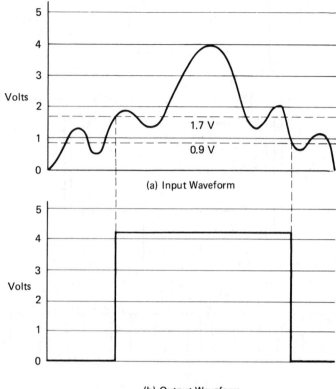

(a) Input Waveform

(b) Output Waveform

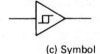

(c) Symbol

FIG. 8-5 Schmitt Trigger Action

within the symbol representing a hysteresis curve, for a Schmitt trigger has electronic hysteresis.

8-3 MULTIPLEXERS, DEMULTIPLEXERS, DECODERS, AND CODE CONVERTERS

Now that's a mouthful! Let's see if we can break that up into smaller bites.

A multiplexer, also called a data selector, is analogous to a single-pole, *n*-throw switch—a selector switch (Fig. 8-6). It selects one of several inputs and then feeds it through to a single output. The CONTROL LINES determine which input is selected.

A demultiplexer is the opposite of a multiplexer; it takes one input and routes it to one of many outputs, depending on the information on its CONTROL LINES (Fig. 8-7). A decoder is a device that converts a coded input (*BCD*, for example) to a single-lead-per-input-word output. A code converter is a device that converts from one code (*BCD*, for example) to another (Gray, for example).

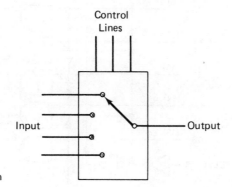

FIG. 8-6 Multiplexer Block Diagram

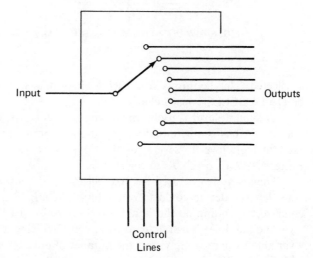

FIG. 8-7 Demultiplexer

Multiplexers

The designer may choose from several available units: 2 inputs to 1, 4 to 1, 8 to 1, and 16 to 1. Figure 8-8 shows a quadruple, 2-to-1-line multiplexer, whereas Fig. 8-9 illustrates a 16-to-1-line device. In each case data are entered via the DATA INPUT leads, the proper input routed to the output by means of the DATA SELECT leads. A STROBE is also provided to allow the designer to disable all output data until a specified time. Then, by allowing the STROBE to go low, the proper lead can

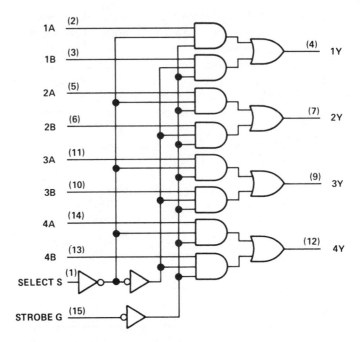

FIG. 8-8 74157 2-to-1 Multiplexer (Courtesy Texas Instruments, Incorporated)

be selected. This feature is very useful where data might be changing the same time the DATA SELECT leads change.

The multiplexer is useful in routing several words to the same destination, such as a bus. (A bus is a wire that accepts inputs from several sources and distributes the resultant logic level to several destinations.) In Fig. 8-10, they are used for routing resisters *A*, *B*, *C*, or all 0's to register *D*.

They can also be used for parallel to serial conversion, as shown in Fig. 8-11. Note that the device could transmit bits *ABCDEFG* in sequence or *GFEDCBA* in sequence, depending on the signals on the DATA SELECT leads.

The multiplexer can be used to directly implement any two-level logic circuit of *N* variables, where *N* represents the number of data select lines. For example, assume the following function must be designed:

$$F = \sum m(0, 3, 5, 6)$$

Figure 8-12 illustrates the technique. When *ABC* is 000, the *I0* input, pin 4, is gated through to the *Z* output, pin 5; when 011, *I3* is gated to *Z*. Thus, each input represents a minterm. By connecting all the inputs shown within the equation (0, 3, 5, and 6) to +5 V and the remaining inputs (1, 2, 4, and 7) to ground, the function is easily implemented.

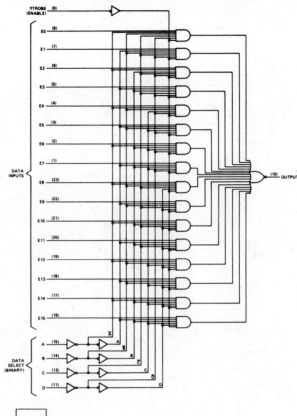

FIG. 8-9 74150 16-to-1 Multiplexer (Courtesy Texas Instruments, Incorporated)

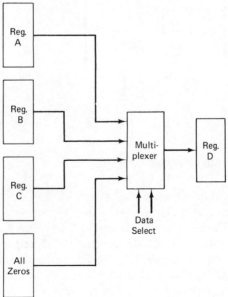

FIG. 8-10 Multiplexing Register Data

245

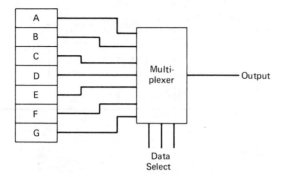

Data
Select

FIG. 8-11 Parallel to Serial Conversion

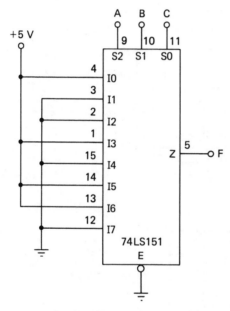

F = Σm (0, 3, 5, 6)

FIG. 8-12 Two-Level Logic Using a Multiplexer

Demultiplexers

Demultiplexers are available with 4, 8, or 16 outputs. The 4- and 16-output devices are shown in Figs. 8-13 and 8-14. Inputs *A* and *B* select which OUTPUT line is active. Data is entered via the DATA line on the 74155 and via one of the *G* leads on the 74154. The STROBE lead (second *G* lead of the 74154) can be used to activate or deactivate the entire IC, allowing time for the address lines to change before information is fed to the output.

Demultiplexers are useful anytime information from one source must be fed several places. For example, assume a 16-bit computer must transmit 48 bits to an

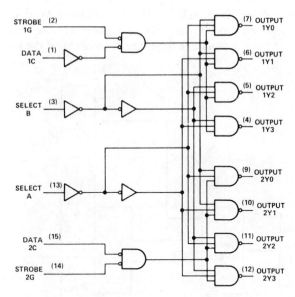

FIG. 8-13 74155 Demultiplexer (Courtesy Texas Instruments, Incorporated)

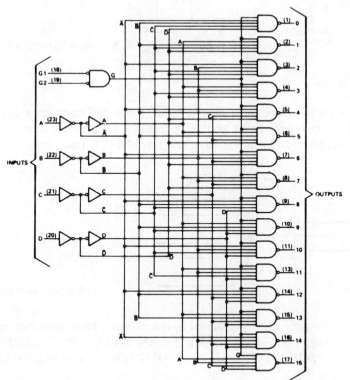

FIG. 8-14 74154 4-Line to 16-Line Demultiplexer (Courtesy Texas Instruments, Incorporated)

output device (Fig. 8-15). The first computer word would appear in the X register and then be clocked into the A register via the demultiplexer. The second word appearing in X would be fed to the B register and the third word in X to the C register.

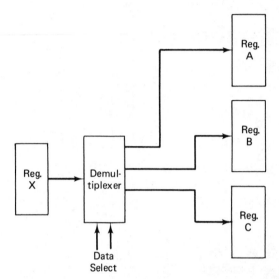

FIG. 8-15 Demultiplexing a Computer Word

Decoders

A decoder is a device that takes a binary word and energizes a particular line based upon the contents of that word. For example, a binary 101 at the input of Fig. 8-16 will cause line 5 to be energized, and a three-bit 011, line 3.

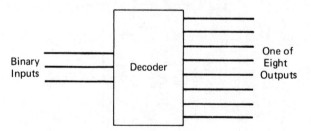

FIG. 8-16 One of Eight Decoder

Any of the demultiplexers shown in the previous paragraph can be used as a binary decoder by providing a continuous 1 on the DATA INPUT leads. For example, a 1011 on the $ABCD$ INPUT leads of the 74154 (4 to 16 line) (Fig. 8-14) will select output 11.

Figures 8-17, 8-18, and 8-19 illustrate BCD to decimal, XS3 to decimal, and XS3 Gray to decimal decoders.

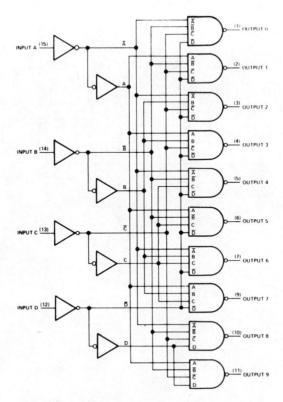

FIG. 8-17 7442A BCD to Decimal Decoder (Courtesy Texas Instruments, Incorporated)

When a decoder is combined with a high-current-driving output it is called a decoder/driver. One such device is a 7447 (Fig. 8-20), used for decoding and driving a seven-segment readout device. Each output lead drives one of the light segments necessary for the readout device. When the 7447 is used in driving light-emitting diode (LED) readouts, each segment requires considerable current. To reduce this load on the power supply, many systems multiplex the driver. Figure 8-21 illustrates such a system to drive five digits. The CONTROL circuit first commands the multiplexer to gate the 3 digit through to the 7447. This IC grounds the proper segments, $a, b, c, d,$ and g of all five LED in order to form the 3. However, the CONTROL circuit selects only the left-hand digit current switch, and only that digit will illuminate. The CONTROL then advances to the 4 digit, selecting the proper multiplexer channel and current switch. This procedure continues until all five have been displayed and then starts over again. The display switches so rapidly that no flicker can be observed.

If this multiplexing method still requires too much current, the CONTROL can supply a STROBE pulse of any length to the 7447.

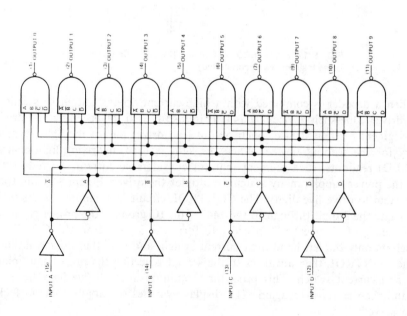

FIG. 8-19 7444A XS3 Gray to Decimal Decoder (Courtesy Texas Instruments, Incorporated)

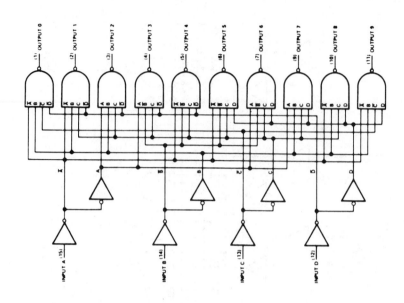

FIG. 8-18 7443A XS3 to Decimal Decoder (Courtesy Texas Instruments, Incorporated)

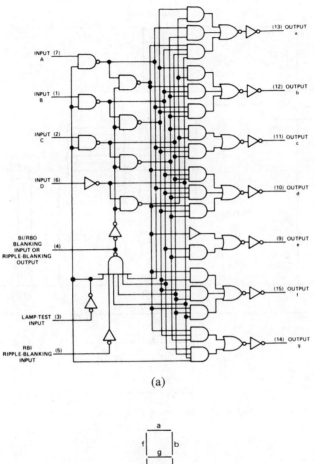

(a)

(b)

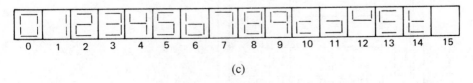

(c)

FIG. 8-20 7447 BCD-to-Seven Segment Decoder/Driver; (a) Logic Diagram; (b) Segment Identification; (c) Numerical Designations and Resultant Displays (Courtesy Texas Instruments, Incorporated)

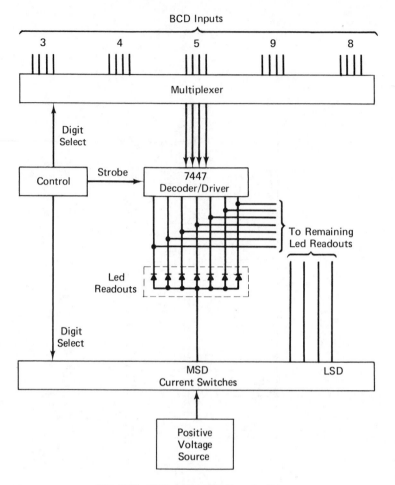

FIG. 8-21 Multiplexing LED Decoder Drivers

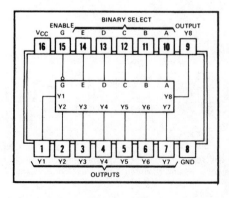

FIG. 8-22 Package Diagram, 74184 and 74185 Code Converters (Courtesy, Texas Instruments Incorporated)

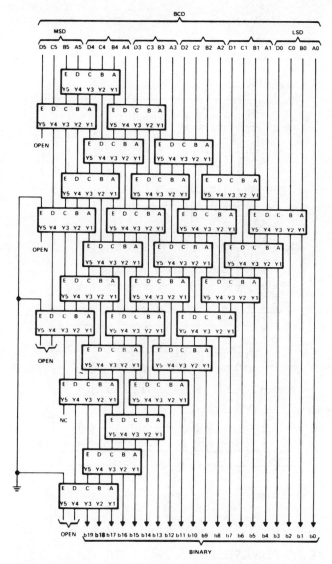

FIG. 8-23 BCD to Binary Converter for Six Decades Using 74184s
(Courtesy Texas Instruments, Incorporated)

Code Converters

A code converter is a device that converts binary signals from a source code (BCD, for example) to its output code (binary, for example). The decoders presented above were also code converters by this definition since they converted signals from one binary system to another.

Two code converters are presented here: the 74184, a BCD to binary converter, and a 74185, a binary to BCD converter (Fig. 8-22). Figure 8-23 illustrates the con-

253

nections necessary to convert from six-decade BCD to binary using 74184s. Figure 8-24 illustrates how to convert from 16-bit binary to five-decade BCD using 74185s.

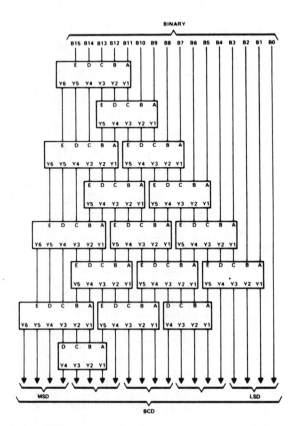

FIG. 8-24 16-Bit Binary to BCD Converter Using 74185A
(Courtesy Texas Instruments, Incorporated)

8-4 ARITHMETIC FUNCTIONS

There are several integrated circuits that perform arithmetic functions: addition, subtraction, multiplication, and division. In this section we shall examine both the theory of these devices and the integrated circuits themselves.

The Half-Adder

The most basic arithmetic circuit is called a half-adder. To understand it better, we shall first examine a binary addition of two numbers [Fig. 8-25(a)]. Here, two one-bit words are added, resulting in two bits of data: a sum bit and a carry bit. Note

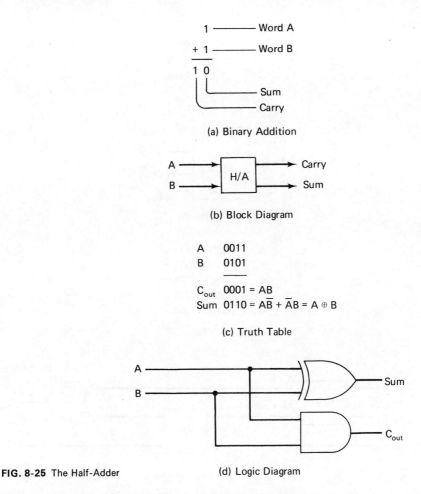

FIG. 8-25 The Half-Adder

that this is a general case: adding two bits of information always results in a sum and a carry. For example, 0 plus 0 results in a sum of 0 and a carry of 0.

Let us design a half-adder. We shall first draw the block diagram [Fig. 8-25(b)]. The device will have two inputs, bits A and B, and two outputs, a sum and a carry. Next, the truth table is constructed. Note that each column is one condition of A and B resulting in a sum and a carry. Also note that the sum is the exclusive **OR** of A and B. Thus, to add is to **XOR**. The final logic diagram is shown in Fig. 8-25(d).

The Full Adder

A full adder is a device capable of adding three binary bits, resulting in a sum and a carry. Consider the binary addition shown in Fig. 8-26(a). Note that the column shown in the box requires addition of three different bits: one bit from word A, one

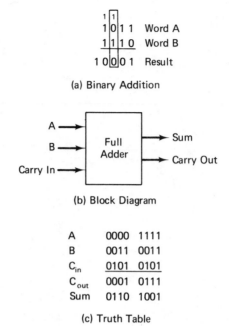

(a) Binary Addition

(b) Block Diagram

(c) Truth Table

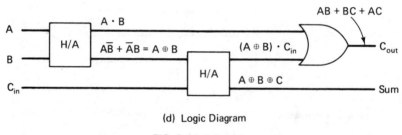

(d) Logic Diagram

FIG. 8-26 Full Adder

bit from word *B*, and a carry in from the previous stage (shown by the small one). A half-adder simply cannot perform this function.

The block diagram for the device is shown in Fig. 8-26(b). Let us now design the circuit. Note that when the truth table is constructed it results in a sum that is the exclusive **OR** of *A*, *B*, and C_{in}. Therefore, this full adder can be constructed of two half-adders using hybrid logic or can be constructed using SOP logic with three input gates. Figure 8-26(d) shows the full adder constructed from two half-adders and an **OR** gate.

In most logic designs, more than one-bit words are to be added. Therefore, although there are one-bit adders available on a chip, most designs use a four-bit adder. These four-bit adders are capable of adding two four-bit words, giving a

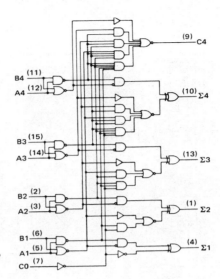

FIG. 8-27 74283 Four-Bit Adder (Courtesy Texas Instruments, Incorporated)

four-bit result and a carry. Figure 8-27 illustrates such an adder, the 74283. Word A is added to word B, along with a carry in ($C0$), resulting in a four-bit sum (Σ) and a carry out ($C4$). Rather than propagating the carry out through four separate adding states, it is generated by two-level logic from the outputs of the $A + B$ and $A \cdot B$ gates. This allows the carry to be generated in the same time or faster than the sum. This type of carry is called a *look-ahead* or *fast* carry and is available as a separate integrated circuit, the 74182.

The 74181 is an arithmetic logic unit capable of adding, subtracting, incrementing, and decrementing and performing shift operations and logical operations (**AND, OR, XOR**) and a variety of other functions. Shown in Fig. 8-28, the four-bit input ($A0$, $A1$, $A2$, $A3$) is combined with a second four-bit input ($B0$, $B1$, $B2$, $B3$) and a carry in (C_n), forming a four-bit output ($F0$, $F1$, $F2$, $F3$) and a carry out (C_{n+4}). The way in which these two words are combined is determined by the control leads ($S0$, $S1$, $S2$, $S3$) and the mode lead (M), as shown in Fig. 8-28(b). Note that when the mode line is high ($M = H$) the unit performs logic functions, whereas when the mode line is low ($M = L$) it performs arithmetic functions. The device can also be operated as a low-true unit.

Any of these four-bit adders can be used as the heart of a BCD adder. In one method, a four-bit BCD digit, word A, is added to another four-bit BCD digit, word B, and a carry in giving a four-bit sum and carry out. This sum is then fed into a second adder where 6 (0110) is added. If the result of this second operation provides a carry out, the answer is taken from this second adder. If the result of this second adder does not provide a carry out, the sum from this stage is taken from the first adder. The carry is taken from the output of the second stage. Similar techniques can be used for XS3.

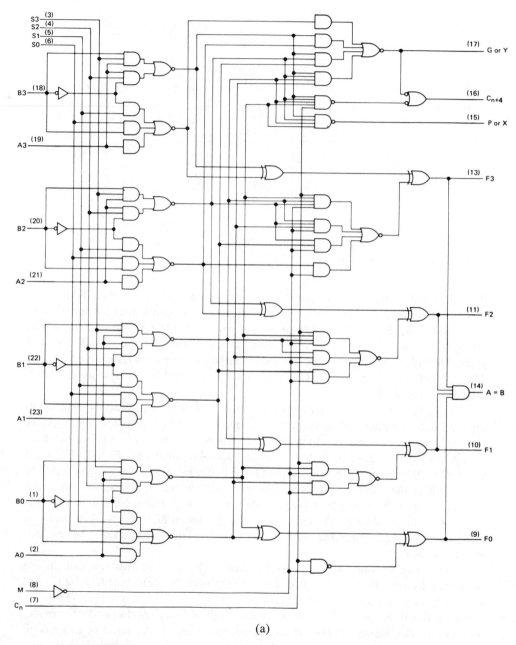

Fig. 8-28 74181 Arithmetic Logic Unit; (a) Logic Diagram (Courtesy Texas Instruments, Incorporated)

SELECTION	ACTIVE-HIGH DATA		
	M = H	M = L; ARITHMETIC OPERATIONS	
	LOGIC	$C_n = H$	$C_n = L$
S3 S2 S1 S0	FUNCTIONS	(no carry)	(with carry)
L L L L	F = $\bar{A}$	F = A	F = A PLUS 1
L L L H	F = $\bar{A}$ + B	F = A + B	F = (A + B) PLUS 1
L L H L	F = $\bar{A}$B	F = A + $\bar{B}$	F = (A + $\bar{B}$) PLUS 1
L L H H	F = 0	F = MINUS 1 (2's COMPL)	F = ZERO
L H L L	F = $\overline{AB}$	F = A PLUS A$\bar{B}$	F = A PLUS A$\bar{B}$ PLUS 1
L H L H	F = $\bar{B}$	F = (A + B) PLUS A$\bar{B}$	F = (A + B) PLUS A$\bar{B}$ PLUS 1
L H H L	F = A $\oplus$ B	F = A MINUS B MINUS 1	F = A MINUS B
L H H H	F = A$\bar{B}$	F = A$\bar{B}$ MINUS 1	F = A$\bar{B}$
H L L L	F = $\bar{A}$ + B	F = A PLUS AB	F = A PLUS AB PLUS 1
H L L H	F = $\overline{A \oplus B}$	F = A PLUS B	F = A PLUS B PLUS 1
H L H L	F = B	F = (A + $\bar{B}$) PLUS AB	F = (A + $\bar{B}$) PLUS AB PLUS 1
H L H H	F = AB	F = AB MINUS 1	F = AB
H H L L	F = 1	F = A PLUS A*	F = A PLUS A PLUS 1
H H L H	F = A + $\bar{B}$	F = (A + B) PLUS A	F = (A + B) PLUS A PLUS 1
H H H L	F = A + B	F = (A + $\bar{B}$) PLUS A	F = (A + $\bar{B}$) PLUS A PLUS 1
H H H H	F = A	F = A MINUS 1	F = A

*Each bit is shifted to the next more significant position.

(b)

SELECTION	ACTIVE-LOW DATA		
	M = H	M = L; ARITHMETIC OPERATIONS	
	LOGIC	$C_n = L$	$C_n = H$
S3 S2 S1 S0	FUNCTIONS	(no carry)	(with carry)
L L L L	F = $\bar{A}$	F = A MINUS 1	F = A
L L L H	F = $\overline{AB}$	F = AB MINUS 1	F = AB
L L H L	F = $\bar{A}$ + B	F = A$\bar{B}$ MINUS 1	F = A$\bar{B}$
L L H H	F = 1	F = MINUS 1 (2's COMP)	F = ZERO
L H L L	F = $\overline{A + B}$	F = A PLUS (A + $\bar{B}$)	F = A PLUS (A + $\bar{B}$) PLUS 1
L H L H	F = $\bar{B}$	F = AB PLUS (A + $\bar{B}$)	F = AB PLUS (A + $\bar{B}$) PLUS 1
L H H L	F = $\overline{A \oplus B}$	F = A MINUS B MINUS 1	F = A MINUS B
L H H H	F = A + $\bar{B}$	F = A + $\bar{B}$	F = (A + $\bar{B}$) PLUS 1
H L L L	F = $\bar{A}$B	F = A PLUS (A + B)	F = A PLUS (A + B) PLUS 1
H L L H	F = A $\oplus$ B	F = A PLUS B	F = A PLUS B PLUS 1
H L H L	F = B	F = A$\bar{B}$ PLUS (A + B)	F = A$\bar{B}$ PLUS (A + B) PLUS 1
H L H H	F = A + B	F = A + B	F = (A + B) PLUS 1
H H L L	F = 0	F = A PLUS A*	F = A PLUS A PLUS 1
H H L H	F = A$\bar{B}$	F = AB PLUS A	F = AB PLUS A PLUS 1
H H H L	F = AB	F = A$\bar{B}$ PLUS A	F = A$\bar{B}$ PLUS A PLUS 1
H H H H	F = A	F = A	F = A PLUS 1

(c)

FIG. 8-28 (continued); (b) Function Table, Active High Data; (c) Function Table, Active Low Data (Courtesy Texas Instruments, Incorporated)

Although the 74181 can be used to determine whether one binary word is identical to a second word, the 7485 (Fig. 8-29) can be used to determine if $A = B$, $A > B$, or $A < B$. ICs can be cascaded to compare magnitudes for any binary word length.

A four-bit multiplier is also available (Fig. 8-30), inputting a four-bit multiplier and a four-bit multiplicand and outputting an eight-bit result. Two ICs are necessary,

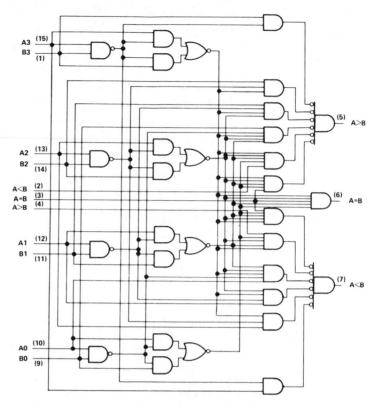

FIG. 8-29 7485 4-Bit Magnitude Comparator (Courtesy Texas Instruments, Incorporated)

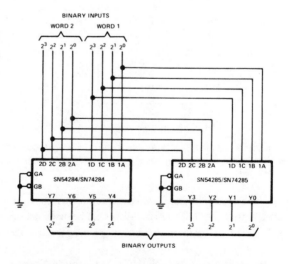

FIG. 8-30 4 By 4 Binary Multiplier (Courtesy Texas Instruments, Incorporated)

one for the low-order four-bit result and one for the high-order four-bit result. Using these ICs and the 74181 Arithmetic Logic Unit, an 8×8 multiplier can be built with a 16-bit result. It should be mentioned that this type of multiplication is much faster, but less prevalent, than the usual shift and add method.

8-5 FLIP-FLOPS

All four types of flip-flops, RS, T, D, and JK, in addition to the monostable multivibrator, are available to the designer. However, before discussing them, let us examine two different types of clocking schemes used with flip-flops: pulse triggered and edge triggered.

Pulse-triggered Flip-flops

In our investigation of flip-flops, we analyzed a simple RS device that could contain a 1 or a 0 (Fig. 8-31). This circuit works well for unclocked devices. However,

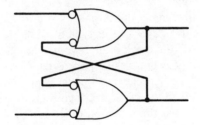

FIG. 8-31 RS Flip-flop

when a clock is included, a different circuit must be used, incorporating at least two flip-flops. This becomes apparent when the device is analyzed as an asynchronous circuit. For example, a type T flip-flop would require the following states:

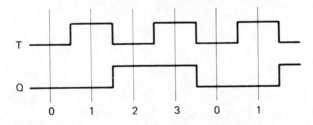

Note that in order to provide these four states two flip-flops are required to make up one type T flip-flop. Similar problems arise for RS, D, and JK FFs with clocks. Each requires at least two internal FFs.

Figure 8-32 illustrates how these two internal FF's are connected to form a JK device. When the clock is high, the information on the J and K leads causes the master FF to go to the correct next state. When the clock goes low, the information

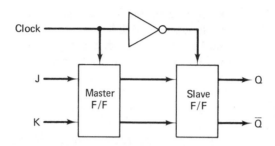

FIG. 8-32 Block Diagram, *JK* Master-slave Flip-flop

in the master is transferred to the slave, and the *J* and *K* inputs are disconnected from the master.

A simplified logic diagram for a 74H71 *JK* pulse-triggered flip-flop is shown in Fig. 8-33. Assume *Q* is low, *J* high, and *K* low. When the clock goes high the *J* gate will **AND**, causing point 1 to go low and point 3 high. Since *K* is low, 2 will be high and 4 will go low, keeping the FF in this state. When the clock goes low, both the *J* and *K* gates will go high, transferring the *EF* information to the *AB* FF such that *Q* will be high. Note that the *J* and *K* gates are both disabled when the clock is low.

One of the disadvantages of this type of FF is that both the *J* and *K* leads must be held constant while the clock is high. For example, assume *Q* high and *J* and *K* low. When the clock goes high, the *J* and *K* gates are disabled. Now assume *K* goes high momentarily. This will cause a low-level pulse on the *K* gate output, causing point 4 to go high. When the clock again goes low, point 6 will go low, causing *Q̄* to go high. Thus, the FF has flipped, even though the *K* lead returned to a low prior to the high-to-low transition of the clock pulse.

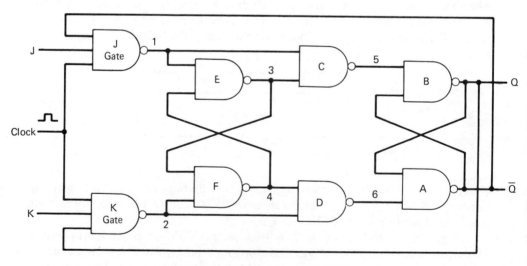

FIG. 8-33 74H71 Pulse-triggered Flip-flop

This type of FF, where the data are entered on the rising edge of the clock pulse and then transferred to the output on the falling edge of the pulse, is called a pulse-triggered FF.

The example we have chosen above requires a positive pulse. By inserting an inverter in series with the clock line, the device will require a negative pulse.

Edge-triggered Flip-flops

The major disadvantage of the pulse-triggered FF is that the J and K leads must be held constant while the clock is active. The edge-triggered FF overcomes this problem by activating on one of the edges (rising or falling) of the clock pulse.

Figure 8-34 is a simplified logic diagram of the 74S112 FF, which triggers on the falling edge of its clock pulse. Assume Q is low, J high, and K low and the clock low. When the clock goes high, gate A output will go low and gate B high. Since $\bar{Q}$ is high, gate C will **AND**, causing Q to remain low. Note that the only result of the clock high is the activation of the AB FF.

When the clock goes low, gate C will be disabled, causing Q to go high. This will cause F to **AND** and $\bar{Q}$ to go low. Note that the action taken by the GH FF depends on what AB was prior to the falling edge of the clock pulse. Also note that data must be entered from the JK leads into the AB FF prior to this time. This prerequisite is called the *setup time*.

Although the edge-triggered device overcomes one problem, it requires a fast-falling edge for the clock pulse. If a clock does not have this sharp edge, it can be

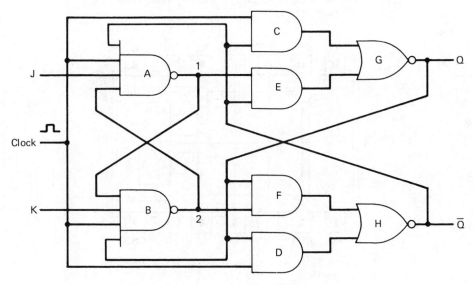

FIG. 8-34 74S112 *JK* Edge-triggered Flip-flop

fed through a Schmitt trigger to square it up. You see, the edge-triggered FF is only for squares.

Type *RS* Flip-flops

Two types of *RS* FF are available: unclocked and clocked. Figure 8-35 illustrates the unclocked type, also called an *RS* latch. Figure 8-36 illustrates a pulse-triggered, master-slave *RS* FF. Note that the *R* and *S* leads cannot both be high at the same time. The device also includes CLR (Clear, low true) and PR (Preset, low true) leads. These inputs take priority over any clock (CK), *R*, or *S* inputs.

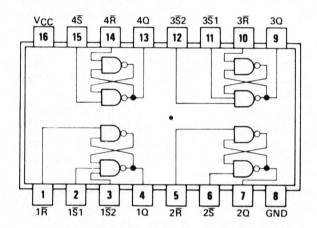

FIG. 8-35 74279 *RS* Latches (Courtesy Texas Instruments, Incorporated)

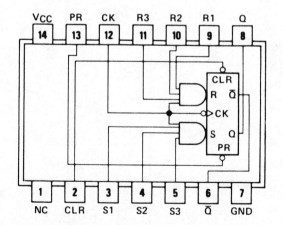

FIG. 8-36 74L71 *RS* Flip-flop (Courtesy Texas Instruments, Incorporated)

Type *T* Flip-flops

Since any *JK* flip-flop can be made into a type *T* by holding *J* and *K* high, type *T* devices are not available as discrete units. They are included as part of larger IC functions, such as ripple counters.

Type *D* Flip-flops

Figure 8-37 illustrates a 7474 type *D* FF with PRESET and CLEAR. This particular one is edge triggered. Eight type *D*s are available in a single package (Fig. 8-38).

Another type of device is called a *DG* FF (*D* for type *D*, *G* for gated). Figure 8-39 illustrates the 7475, in which data are entered into the device and transmitted to its output when *G* is high; when *G* goes low, the input leads are disabled but the FF remains in the state it was prior to *G* going low.

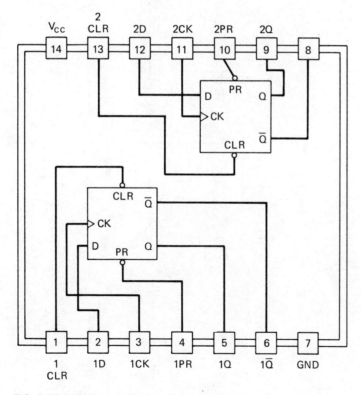

FIG. 8-37 7474 Type *D* Flip-flop (Courtesy Texas Instruments, Incorporated)

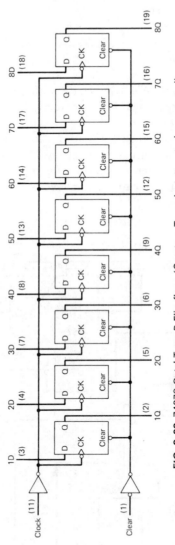

FIG. 8-38 74273 Octal Type *D* Flip-flops (Courtesy Texas Instruments, Incorporated)

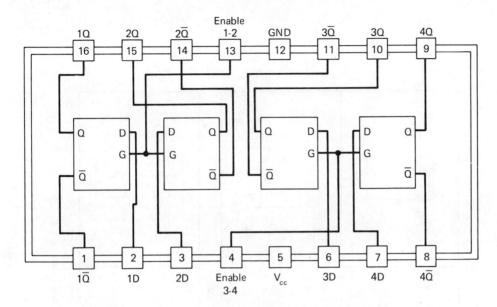

FIG. 8-39 7475 Bistable Latch (Courtesy Texas Instruments, Incorporated)

Type *JK* Flip-flops

Like people, *JK* FF's come in a variety of sizes and shapes. They can be pulse-triggered or edge-triggered on low-to-high or high-to-low transitions; they can be equipped with PRESET and/or CLEAR inputs or equipped with gates on the *J* and *K* input leads. Figure 8-40 illustrates most of the available options. Some *JK*s have a *K̄* lead. This enables the designer to connect the *J* to the *K̄* lead, resulting in a type D FF.

Monostable Multivibrators

A monostable multivibrator, also called a one-shot or single-shot, is a device that outputs a pulse of predetermined length every time the input level changes from a logical 0 to a logical 1. The length of the output pulse is usually determined by a resistor-capacitor network. Figure 8-41 illustrates the device. Each time an input pulse is received, the output provides a 50-ms pulse in this particular circuit.

There are two types of devices available: nonretriggerable and retriggerable. When a nonretriggerable device is stimulated, it outputs its pulse; however, it locks out any further inputs until the full pulse has been outputted. Duty cycles of up to 90% are possible using this single-shot.

A retriggerable multivibrator is capable of being triggered while the output pulse is being generated, extending the length of the pulse. By continually pulsing its input, the output pulse can be lengthened indefinitely. One use for this capability is to determine if a pulse is missing in a sequence. By selecting the timing values to be

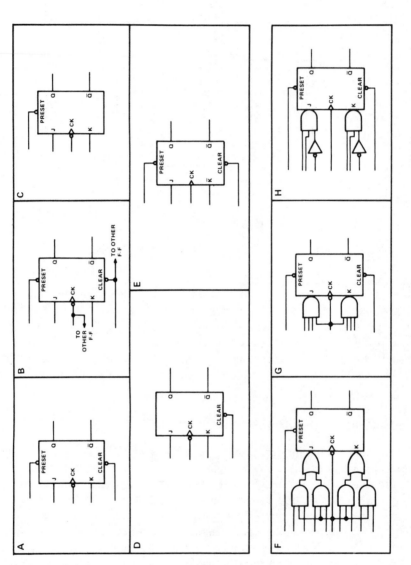

FIG. 8-40 *J/K* Flip-flops (Courtesy Texas Instruments, Incorporated)

268

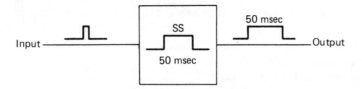

FIG. 8-41 Monostable Multivibrator

greater than the pulse period, the device would be constantly retriggered, until one pulse came up missing. At that time the *SS* (single-shot) would run out.

Sometimes, SSs have a CLEAR input. This allows the device to be cleared prior to the SS pulse running out.

Figure 8-42 illustrates two SSs. The 74121 is nonretriggerable and has a Schmitt

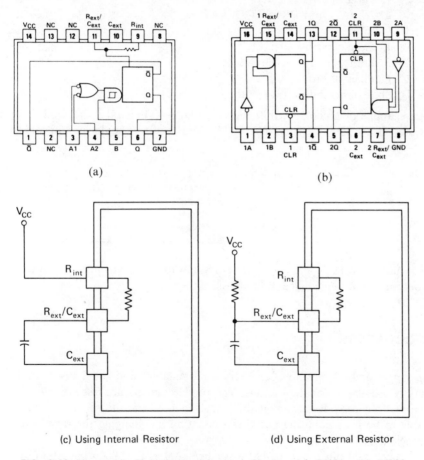

(a)

(b)

(c) Using Internal Resistor

(d) Using External Resistor

FIG. 8-42 Monostable Multivibrator Integrated Circuits; (a) 74121; (b) 74123 (Courtesy Texas Instruments, Incorporated); (c) Using Internal Resistor; (d) Using External Resistor (Courtesy Texas Instruments, Incorporated)

trigger input. The 74123 is a dual retriggerable type. The 74121 has an internal resistor that can be used by connecting as shown in Fig. 8-42(c). If greater repeatability is required, an external resistor can be provided (in addition to the capacitor), as shown in Fig. 8-42(d). R_{int} should be left open. Either monostable can be made variable by providing a variable external resistor.

8-6 SHIFT REGISTERS

Shift registers are sold in a wide variety of configurations. There are many, many types available. Some of their important options include:

- A. Input:
    1. Parallel, serial, or both.
    2. $JK, J\bar{K}, D$, or PRESET/CLEAR.
- B. Output:
    1. Parallel, serial, or both.
    2. Both Q and $\bar{Q}$ accessible.
- C. Number of bits.
- D. Shift right, shift left, or both.

As an example of a complex shift register, Fig. 8-43 illustrates the 74S299 eight bit universal shift/storage register. The $S0$, $S1$ leads control the device's operation:

S1	S0	Function
0	0	Store (no loading or shifting)
0	1	Shift right
1	0	Shift left
1	1	Synchronous parallel load

The output controls, $\bar{G}_1$ and $\bar{G}_2$, must both be low to obtain an output. If either is high, the output is high impedance (tri-state output).

8-7 COUNTERS

Both ripple and synchronous counters are available in decade, divide-by-12, and binary (divide-by-16) configurations. The ripple counters are less complex (and less expensive) but provide unwanted spikes during state transitions. Some counters are capable of both up counting and down counting by changing the logic level on an input lead.

Figure 8-44 illustrates a 7490A decade/counter, a 7492A divide-by-12 counter, and a 7493 binary counter. All three are ripple devices and have clear inputs. Figure 8-45 illustrates a synchronous, up/down, parallel-loaded decade counter. The DOWN/

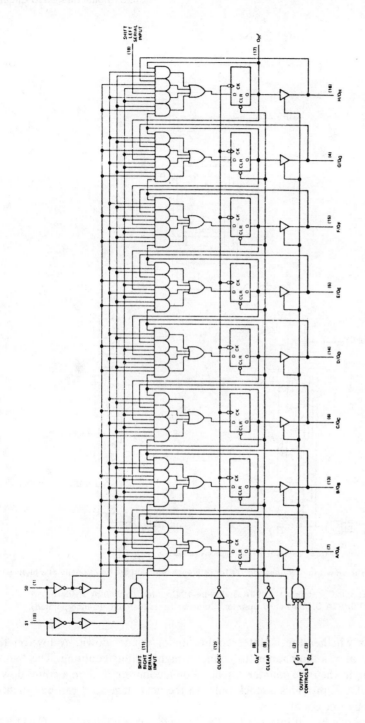

FIG. 8-43 74S299 Shift Register (Courtesy Texas Instruments, Incorporated)

271

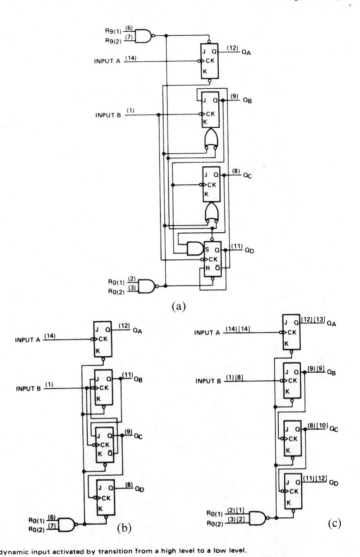

FIG. 8-44 Ripple Counters; (a) 7490A Divide-by-10 Counter; (b) 9492A Divide-by-12 Counter; (c) 7493A Divide-by-16 Counter (Courtesy Texas Instruments, Incorporated)

UP lead controls whether the counter counts up or counts down. A low on the ENABLE G lead allows the counter to count; a high inhibits counting. The MAX/MIN lead goes high when the counter reaches 9 on a count up or 0 on a count down. The RIPPLE CLOCK provides a clock pulse to the next stage so it can be synchronously increased by one count.

Parallel data can be fed into the DATA LEADS while holding the LOAD lead low. This jam-sets/clears each stage of the counter.

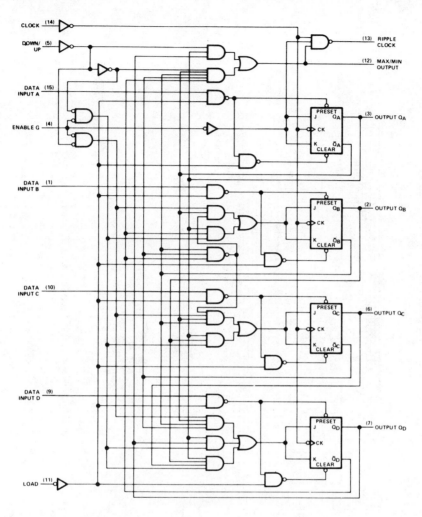

. . . Dynamic input activated by a transition from a high level to a low level.

FIG. 8-45. (Courtesy Texas Instruments, Incorporated)

Rate Multipliers

A rate multiplier is a black box that divides its input clock by a rate that is dependent on a binary input word. Figure 8-46 is a logic diagram of the 7497 binary rate multiplier. The unit's output frequency is

$$F_o = \frac{MF_i}{64}$$

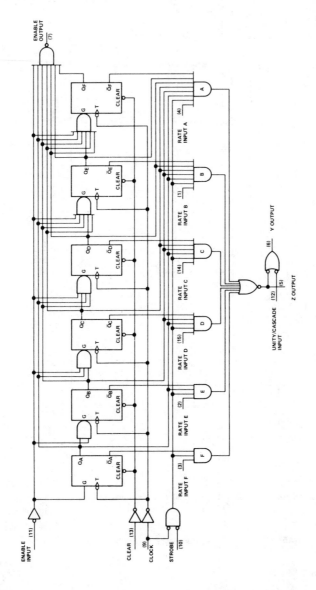

FIG. 8-46 7497 Binary Rate Multiplier (Courtesy Texas Instruments, Incorporated)

where M is a binary word formed from the input leads ($FEDCBA$) and F_i is the input frequency. With an input frequency of 64 kHz and $FEDCBA$ a 010110, the output frequency is

$$F_o = \frac{MF_i}{64} = \frac{010110_2 \cdot 64000_{10}}{64_{10}} = \frac{22 \cdot 64000}{64} = 22 \text{ kHz}$$

The ENABLE INPUT and ENABLE OUTPUT leads connect to previous and next stages, providing for 12- or 18-bit rate multiplication. The STROBE input allows the output (Z output) to be activated.

The 7497 is a binary rate multiplier. By contrast, the 74167 is a four-bit decade rate multiplier obeying the equation

$$F_o = \frac{MF_i}{10}$$

where M is a decimal 0 through 9.

8-8 OPTOELECTRONIC DISPLAY DEVICES

Several devices are available for displaying numbers or letters using light-emitting diodes (LEDs). One such seven-segment device is capable of displaying all the digits 0 through 9 and is available in two configurations, common anode and common cathode. The anode of the common anode device would be connected to $+5$ V and each segment driven by a low-true decoder driver (Fig. 8-21), whereas the cathode of the common cathode device would be connected to ground and each segment driven by a high-true decoder driver.

A second type of device is capable of displaying any character—letters, numbers, and punctuation—by lighting discrete LEDs in a 5×7 dot matrix (Fig. 8-47). The read only memory receives ASCII data to select a character and signals from the ring counter to select the column of diodes being energized. Each character requires five ROM locations, one for each column of diodes. The ring counter then simultaneously selects the ROM location and energizes the correct column, proceeding one at a time until all columns have been scanned. This is done so rapidly that no flicker can be observed.

Other types of LEDs are available for displaying hexadecimal characters (0–9 and A–F). Some units display up to six digits per integrated circuit.

8-9 SUMMARY

Although in Chapters 1 through 7 we explained how to use **AND, OR, NAND,** and **NOR** gates and FF's to design logic systems, using just these elements will result in reinventing the wheel. There are many medium and large-scale integrated circuits that can be used to simplify design and reduce package count. Even logic gates come

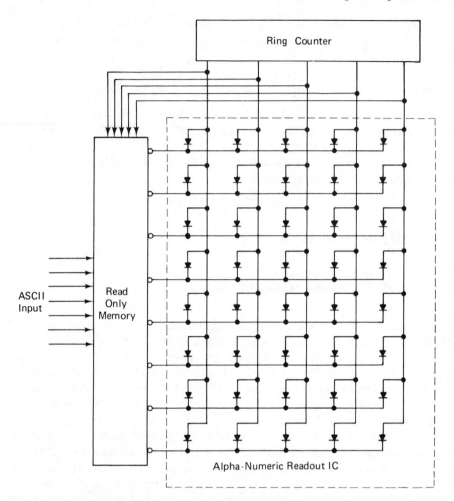

FIG. 8-47 Simplified Diagram, Alphanumeric Display

in a variety of output drive configurations. Schmitt triggers provide hysteresis to "square up" a pulse train. Multiplexers allow several inputs to feed one output, whereas demultiplexers feed one input to several outputs. Decoders and code converters change one type of binary code to another type. Arithmetic chips add, subtract, shift, **OR**, **AND**, and exclusive **OR** binary words.

Pulse-triggered FF's use master/slave configurations and require the inputs to remain constant while the clock is active. Edge-triggered FF's also use master/slave configurations but require a setup time and a sharp clock pulse edge. Monostable multivibrators provide a single pulse out upon receipt of a level transition.

Shift registers can require parallel or serial input, output parallel or serial data, and shift right or left. Counters are available in modulo 10, 12, or 16. Rate multipliers output pulses whose frequency is dependent on an input binary word.

8-10 PROBLEMS

8-1. The outputs of several logic gates are to be wired together. What two types of gate output configurations may be used?

8-2. A logic function, $X = \overline{AB + CD}$, is to be implemented. What IC will provide the smallest package count?

8-3. What device is used to "square up" a logic pulse?

8-4. Four input registers must feed one storage register. What device will perform the 4-to-1 conversion?

8-5. Three control lines must turn on eight different valves, depending on the binary word on the control lines. What device will perform the 3-to-8 conversion?

8-6. A device must convert a four-bit BCD code to the seven-segment code necessary to drive a LED readout. What device will perform the conversion?

8-7. What device can add, **AND**, and shift left?

8-8. Which type of FF can receive slow rise times and fall times on the clock input?

8-9. A clocked *RS* FF has 1's on the *S* and *R* leads. What will the output be upon receipt of the next clock pulse?

8-10. Connecting the *J* and $\bar{K}$ leads of a $J\bar{K}$ FF together results in it acting as a_____.

8-11. A certain clock is continually monitored by a circuit that will go high any time the clock ceases outputting pulses. What type of IC will perform this monitor function?

8-12. What type counter outputs spikes during logic transitions?

8-13. The input frequency of a 7497 binary rate multiplier is 64 kHz. What will its output be if the multiplier word is 1011?

MEMORY
CIRCUITS
AND SYSTEMS

9

9-1 INTRODUCTION TO MEMORIES

We have learned that flip-flops are capable of storing logical 1's and 0's. Thus, they can remember; if we leave the room and return, the same number will be contained within the FF. We now turn our attention to memories capable of storing many 1's and 0's. These devices can be thought of as many FF's contained on a single IC.

Memory Organization

Semiconductor memories are arranged into words, and each word exists at a particular address. Assume we were to visit Stonehill Road in Chelmsford, Massachusetts. At 1 Stonehill Road, there may be a person with a social security number (SSN) of 256-83-4201. At 2 Stonehill Road lives a person with an SSN of 982-46-3512. Thus, associated with each address there is an SSN. In an identical manner, each address of memory contains a particular number, its data word (Fig. 9-1). This memory contains 4096 words at hexadecimal addresses 000 to FFF, and each word contains 8 bits. Thus, it is referred to as a 4096 × 8 bit memory or 4k × 8, for short. Note that the memory address and memory data are not related. Address 000 contains a 0001 1101, whereas address 002 contains a 0111 1101. In a similar manner, a person's home address does not testify to what his SSN is.

Address and Data Buses

Figure 9-2 illustrates the actual signal leads that are fed to and from the memory. Since there are 4096 different address locations, 12 different address leads are required: $2^{12} = 4096$. By putting 0110 1011 0001 on address leads A_{11}-A_0, address 6B1 is fed to

000	0001 1101
001	1101 1101
002	0111 1101
003	0000 1110
004	0010 0100
005	1001 1110
FFD	0011 1100
FFE	1101 1011
FFF	0001 1110

FIG. 9-1 Memory Organization

the memory. Note that A_0 is the least significant bit and A_{11} is the most significant bit. These 12 address leads are called an address bus. Information on the bus is always in parallel; each lead of the bus services a particular bit. A_5, for example, always contains bit 5 of the address being fed to the IC.

The output of this memory is an eight-bit data bus. This bus contains the data at the location addressed by the address bus. Bit 0 of the data bus is always the least significant bit of the data word being addressed by the address bus. For example, in Fig. 9-1, if the address bus had an FFD, the data bus would have 0011 1100 and D_0 would be a 0.

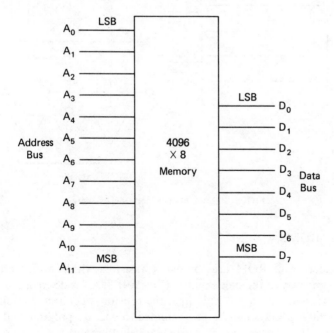

FIG. 9-2 Memory Address and Data Buses

9-2 READ ONLY MEMORIES

The memory chip we have been playing with assumed that the information at each of the addresses was programmed into it and all we wanted to do was read the memory. Such a memory is called a read only memory (ROM).

Information in a read only memory is stored in one-bit elements called cells. Each cell is capable of storing a 1 or a 0. A ROM can be formed from a matrix of diodes as shown in Fig. 9-3, where the input lines, A_0 and A_1, form the address for the memory. Thus this example has a two-bit address bus. For example, if A_1A_0 is 01, word 1 is selected, its NAND gate goes low. This will pull D_2 and D_1 low and leave D_0 high, which, when inverted, will result in a 110 at the output data bus. Thus, memory location 1 contains a 6.

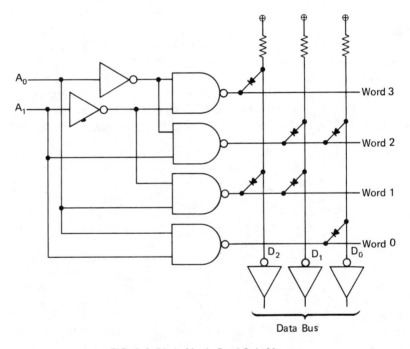

FIG. 9-3 Diode Matrix Read Only Memory

Bipolar ROMs

A second type of ROM uses bipolar transistors within an integrated circuit, much in the same way as the diode matrix (Fig. 9-4). If a 1 is desired, the resistor of a cell is open-circuited, and if a 0 is desired, the resistor is connected as shown. There are two ways these ROMs can be programmed. Some are programmed at the time of manufacture. If a 1 is desired in a particular cell, the resistor is just not connected.

280

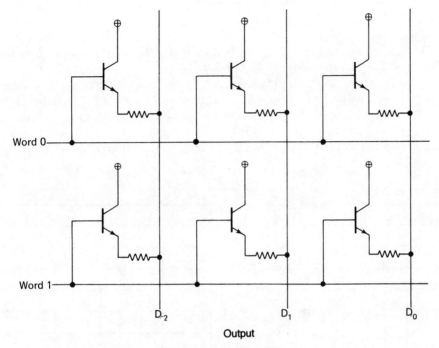

FIG. 9-4 Bipolar ROM Cells

Other ROMs permit the user to program his or her own chip. Each resistor on the IC is manufactured as a fuse so that if the user supplies sufficient current, it can be blown, open-circuiting the emitter of that cell. These are called *programmable read only memories* (PROMs), and the cell type is called a *fuse cell*.

Figure 9-5 illustrates the 74S288 bipolar programmable read only memory (PROM). Pins 10-14 form the address bus—5 bits for 32 locations. Pin 15 is an enable lead and when it is low, the outputs D_0-D_7 will read the memory location addressed

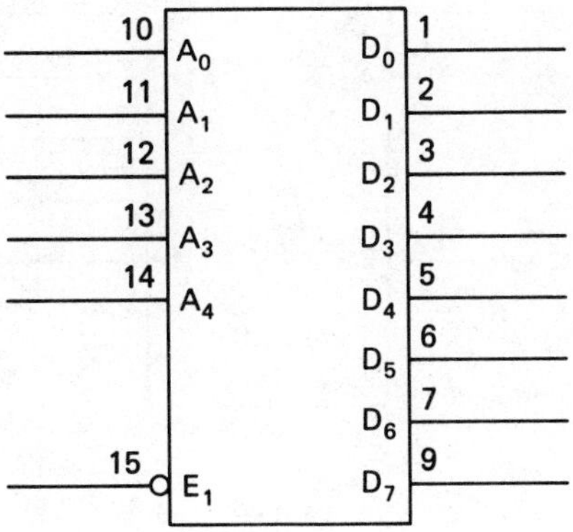

FIG. 9-5 74S288 Bipolar PROM

by the address bus. However, when pin 15 is high, the outputs are at a high impedance level. Using this feature, several IC's may be connected to form a larger memory;
Fig. 9-6 illustrates such a configuration. Note that when pin 15 of $U1$ is high, pin 15
of $U2$ is low. In this manner, $U1$ is the bottom half of the 64 x 8 memory (locations
0-31) and $U2$ is the top half (locations 32-63).

There are two critical timing delays that must be considered when using any
ROM. The first is the delay it takes from when the address is placed on the address
bus to the time the correct data appears at the data bus, assuming the chip is selected
($\overline{E1}$ is low). The second timing delay is that from the time the chip is selected by $\overline{E1}$
going low to the time the data is valid at the output, assuming the address bus is

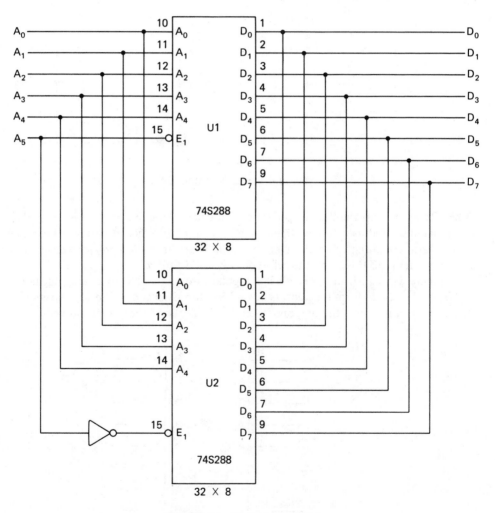

FIG. 9-6 Forming a 64 × 8 PROM

stable. It should be noted that, until these two timing delays are met, the outputs will be displaying invalid and unusable data.

MOS ROMs

A third type of ROM is formed from a MOS IC (Fig. 9-7). In this cell, the silicon gate is completely insulated from the N-type substrate. However, if a sufficiently high voltage is applied to the source or drain, the PN junction avalanches, injecting high-energy electrons into the insulated silicon gate. Since the gate is completely surrounded by insulator material, it cannot leak very rapidly; it loses about 30% of its charge in 10 years.

These MOS ROMs can be programmed in two ways. One is by placing the charge on the gate at the time of manufacture. However, some devices are manufactured with a quartz window in the IC package, allowing the charge to be removed under high-intensity ultraviolet light. This is caused by the flow of photo current from the gate to the substrate. Having been returned to its original neutral charge, the gate may now be reprogrammed by applying sufficient avalanche voltage. This erasable feature allows the designer to thoroughly shake down the program before committing it to the mask procedure.

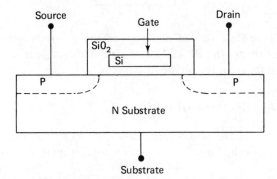

FIG. 9-7 MOS Read Only Memory

Figure 9-8 illustrates the 2764 erasable, programmable, read only memory (EPROM). The IC is read by causing pins 2, 22, and 27 (the chip select leads) all to go low, and supplying an address to the address bus, A_0-A_{11}. Data will appear at the data output, D_0-D_7 (after a suitable propagation delay, of course). When any of the chip selects are high, the output leads will "float" at a high impedance; this feature allows paralleling several chips to form an even larger memory.

Programming is done by:

A. Causing pins 2 and 27 to go low and pin 22 to go high.
B. Supplying an address to the address bus.
C. Placing the data to be written onto the data bus.
D. Supplying 25 V to pin 1. (For normal read and write, pin 1 is at 5 V.)
E. Pulsing pin 22 from $+5$ V to 0 V.

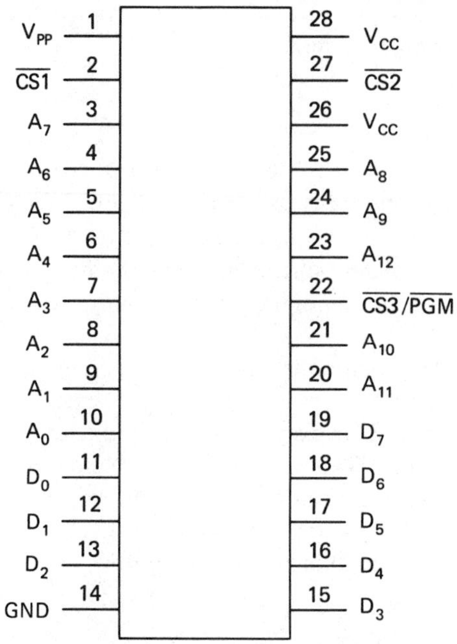

FIG. 9-8 2764 EPROM

Most manufacturers of digital products buy a programming machine that does all this for them. The EPROM is erased by exposing its quartz window to a 12 mW/cm² ultraviolet lamp for about 20 minutes.

Applications of ROM

ROMs are used for a variety of tasks within a digital system. They can be used as a direct substitute for any random logic: that is, **AND** gates, **OR** gates, and inverters.

Example 9-1: Use a 32 × 8 bipolar PROM to form the following functions:

$$f_1 = \sum m(0, 2, 5, 6, 8)$$
$$f_2 = \sum m(4, 5, 6, 8, 9)$$
$$f_3 = \sum m(6, 7, 8, 10, 13, 17, 22, 23)$$

Solution:

Since we need only three outputs, we shall assign the output data bits as: $f_1 = D_0, f_2 = D_1, f_3 = D_2$. The remaining output data bits will be left open. Next, make a list of all the locations of the PROM (Fig. 9-9). Each minterm of the expressions represents its own address. Minterm m_4, for example, represents address 4 on the chart. Next, where a minterm is present within an expression, place a 1 in the table corresponding to that bit. Expression f_2, for example, includes minterm 5. Therefore, place a 1 in bit D_1 of address 5. Where

Location	Contents	Location	Contents
00	0000 0001	16	0000 0000
01	0000 0000	17	0000 0100
02	0000 0001	18	0000 0000
03	0000 0000	19	0000 0000
04	0000 0010	20	0000 0000
05	0000 0011	21	0000 0000
06	0000 0111	22	0000 0100
07	0000 0100	23	0000 0100
08	0000 0111	24	0000 0000
09	0000 0010	25	0000 0000
10	0000 0100	26	0000 0000
11	0000 0000	27	0000 0000
12	0000 0000	28	0000 0000
13	0000 0100	29	0000 0000
14	0000 0000	30	0000 0000
15	0000 0000	31	0000 0000

(a) Table

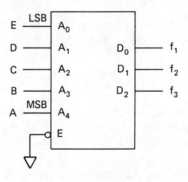

FIG. 9-9 Using PROMs for
Random Logic, Example 9-1   (b) Logic Diagram

a minterm is not present, place a zero within the corresponding data word.
Figure 9-9 shows the completed table and the resulting logic diagram.

ROMs used for direct replacement of random logic provide the advantage of
fewer ICs and ease of circuit modification (just program a new ROM). However,
they have the disadvantages of outputting glitches whenever any of the address lines
change state, and they usually have higher propagation delay than two level logic.

ROMs can also be used for code translation (Fig. 9-10). If ASCII data is placed
on the address bus, the data bus will contain the EBCDIC representation. ROMs

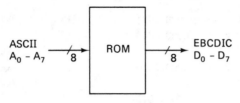

(a) ASCI to EBCDIC

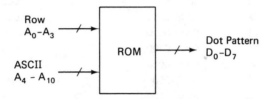

(b) Character Generation

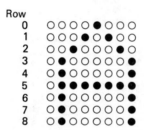

(c) Dot Pattern in the ROM   **FIG. 9-10** Code Translation

can also translate an ASCII character and row number into a dot pattern [Fig. 9-10(b) and (c)]. Address lines A_4-A_{10} select the character and address lines A_0-A_3 select the row of dots that is to be displayed. Most data terminals use this method of generating a character on the video screen.

PROMs, especially EPROMs, are used most extensively to store microcomputer programs. In this way, the programs can be easily changed to suit the purpose of the customer. We shall discuss microcomputers in Chapter 10.

9-3 STATIC RANDOM-ACCESS MEMORIES

A random-access memory (RAM) is a memory that can be both written into and read from. The term, "random access," can be confusing, for it is, technically, any memory in which each memory location can be addressed directly, without having to address another location first. As an example, a circulating shift register is not a random-access memory, for in order to look at a particular bit location, it is necessary to cause the register to circulate until that bit can be accessed. The ROMs and PROMs in the previous paragraphs are, according to this definition, RAMs, for any memory location can be accessed directly. However, although this is the strict definition of a RAM, the term is usually applied only to read/write random-access devices, thus excluding ROMs and PROMs. We shall use the term in this latter sense, for that is the way industry uses it.

Reading and Writing

Two typical RAMs are shown in Fig. 9-11, in each case a 16×1 memory: 16 locations of 1 bit each. Addresses A_0-A_3 select the memory locations. Note that

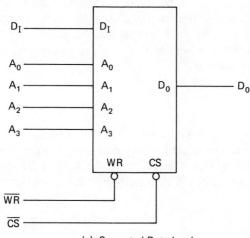

(a) Separated Data Leads

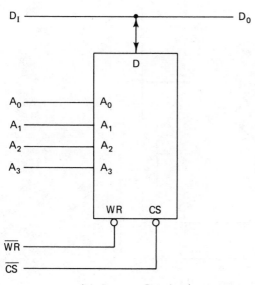

(b) Common Data Lead   **FIG. 9-11** Random-Access Memory (RAM) memory (a) uses one lead, D_I (DATA IN), for input data and one lead, D_O (DATA OUT), for output data whereas memory (b) uses the same lead for DATA IN and for DATA OUT. Reading is accomplished by placing an address onto the address bus, causing $\overline{CS}$ to go low, and keeping $\overline{WR}$ ($\overline{WRITE}$) high. Data will appear at D_O in memory (a) or at D in memory (b). Writing is accomplished by placing an address onto the address bus, causing $\overline{CS}$ to go low, placing the data on D_I or D, then causing $\overline{WR}$ to go low. One note of caution should be observed, however. There are several times at which writing can take place, depending on the particular IC selected:

A. Writing occurs when $\overline{WR}$ goes from high to low or $\overline{CS}$ goes from high low, whichever occurs last. The address must be stable during this transition.
B. Writing occurs when $\overline{WR}$ goes low to high or $\overline{CS}$ goes low to high, whichever occurs first. The address must be stable during this transition.
C. Writing occurs when $\overline{CS}$ and $\overline{WR}$ are both low. The address must remain stable during the entire writing interval.

Bipolar RAMs

The bipolar memory cell is shown in Fig. 9-12. The SENSE, X SELECT, and Y SELECT leads provide a low resistance to ground, effectively grounding the emitters and making the cell a simple flip-flop. To read from the cell, both the X SELECT and the Y SELECT lines must be high; these leads will then have no current flow since

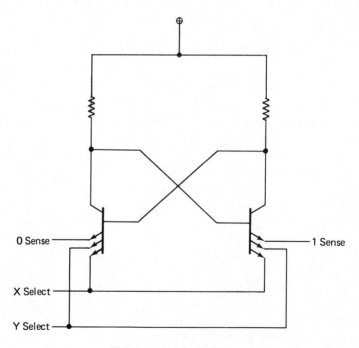

FIG. 9-12 Bipolar RAM Cell

the emitters they feed are reverse biased. The current in each sense lead will then be measured and if present, will represent a logical one. Thus, if 0 SENSE has current and 1 SENSE has no current, the FF is in the zero state.

Data can also be written into the cell. By holding X and Y SELECT lines high, one of the sense leads is held high and the other low. This will turn one transistor on and the other off. When X and Y are returned low the FF will latch in that state.

Figure 9-13 illustrates a 7489 16-word, 4-bit RAM. Data can be written into the memory via the DATA INPUTS by supplying an address to the SELECT inputs

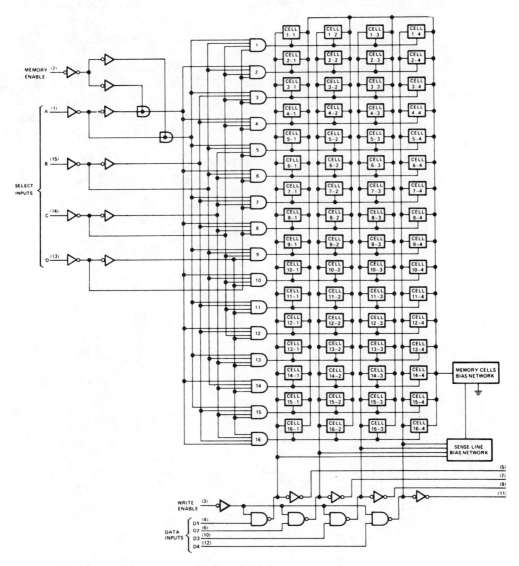

FIG. 9-13 7489 Random-Access Memory (Courtesy Texas Instruments, Incorporated)

and providing low levels on the MEMORY ENABLE and WRITE ENABLE leads.

Data can be read from any memory location by supplying an address to the SELECT inputs, providing a high on the WRITE ENABLE line, and a low on the MEMORY ENABLE. The data will appear in complemented form at the SENSE outputs. Chips can be wired in parallel to increase memory size.

MOS RAMs

There are two basic types of MOS RAMs: static and dynamic. The static RAM uses the cell shown in Fig. 9-14. The 0 SENSE and 1 SENSE lines are used to sense the state of the FF formed by Q_3 and Q_5. Each word must be selected via a WORD SELECT line; this requires more decoding than an X-Y select scheme, but each cell requires only six transistors, compared to eight for an X-Y MOS cell. The FF can be set to the 1 or 0 state by using 0 SENSE and 1 SENSE lines as data input (0 SENSE = $\overline{1\ SENSE}$), and selecting the word to be written into via the WORD SELECT line.

This type of cell is called a static cell, for information remains stored until changed by a new WRITE operation. This is true for all FF's; information remains available indefinitely, unless power is interrupted.

Figure 9-15 illustrates the 4k × 1 2147 MOS RAM. It uses separate DATA IN and DATA OUT leads, and writing occurs whenever $\overline{CS}$ and $\overline{WR}$ are both low.

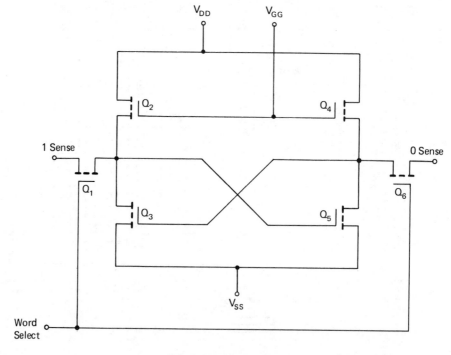

FIG. 9-14 MOS RAM Cell

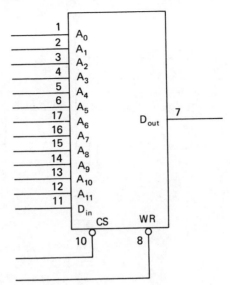

FIG. 9-15 2147 4k × 1 Static MOS RAM

During a write, the D_0 lead is at a high impedance level. This RAM also has a power down feature: when $\overline{CS}$ is high, the chip consumes 12% of the supply current it consumes when $\overline{CS}$ is low. By causing $\overline{CS}$ to go low only during read or write cycles, the designer can cause the memory to consume much less power, saving in cost of the power supply and increasing the reliability.

9-4 DYNAMIC RANDOM-ACCESS MEMORIES

The dynamic memory is one which stores its information on capacitances internal to the chip. Figure 9-16 shows the cell used for this type of memory. In this circuit, the CHIP ENABLE, when not true, turns Q_1 on allowing the SENSE line to be charged to V volts. When CHIP ENABLE goes true, Q_1 turns off, but the SENSE line remains charged because of its internal capacitance, C_s. When a cell is selected, the capacitance of the cell, C_c, and that of the SENSE line, C_s, form a voltage divider. The SENSE AMP LATCH detects this small change in voltage and latches to the 1 or 0 state that it has detected, pulling the SENSE line firmly to the proper high or low voltage, recharging C_c. Note that merely reading a cell will refresh the voltage across C_c. Writing is done by setting or resetting the SENSE AMP LATCH and then selecting the proper cell.

Because the charge on C_c tends to leak off, these cells must be periodically refreshed by reading or writing each cell. This is usually done by a separate circuit, with each cell being refreshed at least once every 4 milliseconds (ms).

Figure 9-17 illustrates the 65k × 1 4164 dynamic RAM (DRAM). The first impression one has when looking at such a chip is, "How can we address 65,536 different memory locations using only 8 address lines? After all, $2^8 = 256$ and $2^{16} =$

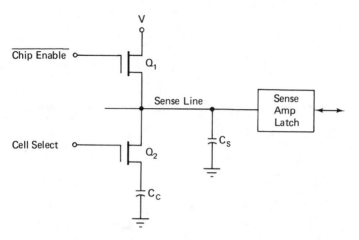

FIG. 9-16 MOS Dynamic RAM Cell

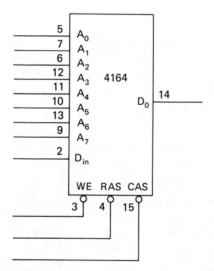

FIG. 9-17 4164 Dynamic RAM

65,536." The solution can be found by examining Fig. 9-18. The memory is arranged into 256 rows and 256 columns. To select a particular memory location, place 8 bits of address onto A_0-A_7 and cause $\overline{RAS}$ (ROW ADDRESS STROBE) to go high to low. This strobes (clocks) 8 bits of address into the ROW REGISTER, selecting one particular row of address. Next, place a column address onto A_0-A_7 and cause $\overline{CAS}$ (COLUMN ADDRESS STROBE) to go high to low. This strobes 8 bits of address into the COLUMN REGISTER, selecting a particular column. We have now selected one of 256 rows and one of 256 columns or one of $256^2 = 65,536$ locations.

Figure 9-19 illustrates how this addressing can be done using a multiplexer. First, address EA_0-EA_7 (EXTERNAL ADDRESS) are selected and outputted to the

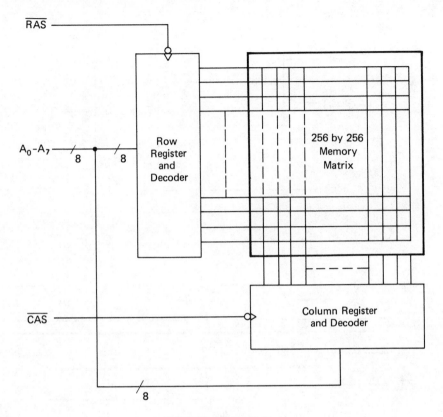

FIG. 9-18 DRAM Matrix

DRAM. $\overline{\text{RAS}}$ is then brought high to low. Next, timing signals switch the multiplexer to select EA_8-EA_{15}, feeding this to the DRAM. Then $\overline{\text{CAS}}$ is brought high to low. If $\overline{\text{WR}}$ remains high, the data will appear at D_0 (after a suitable propagation delay) [Fig. 9-19(b)].

There are two ways in which writing can be done. An early write cycle brings $\overline{\text{WR}}$ low before $\overline{\text{CAS}}$; writing takes place when $\overline{\text{CAS}}$ goes from high to low [Fig. 9-19(c)]. A late write cycle has $\overline{\text{CAS}}$ going low before $\overline{\text{WR}}$ and writing takes place on the high to low transition of $\overline{\text{WR}}$ [Fig. 9-19(d)].

Refresh must be done every 4 ms (minimum) for each row; refreshing a row refreshes all locations associated with that row. This is done by presenting an address to the chip and causing $\overline{\text{RAS}}$ to go from high to low. Each of the 256 row addresses must be presented to the chip and $\overline{\text{RAS}}$ brought high to low during a 4 ms period. Note that $\overline{\text{CAS}}$ is not needed; during a refresh cycle it is usually held high, resulting in D_0 being at a high impedance. Some microprocessors automatically take care of this chore, providing a refresh signal to the memory indicating that a refresh cycle is in progress. Where this is not done, a refresh arbitrating circuit can be purchased. Some memory chips contain such a circuit and are known as pseudo-static memories.

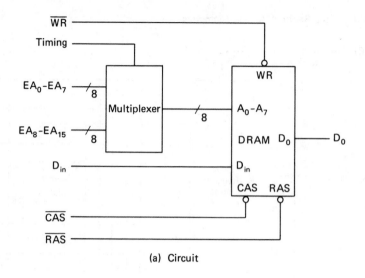

(a) Circuit

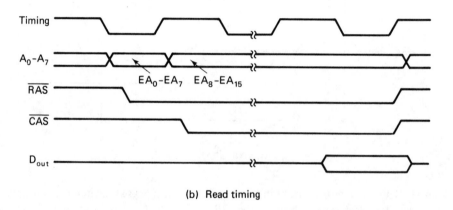

(b) Read timing

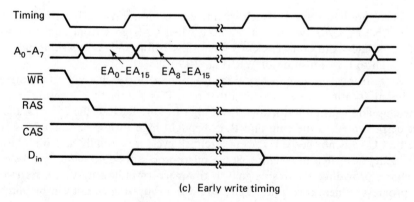

(c) Early write timing

FIG. 9-19 Using the Dynamic RAM

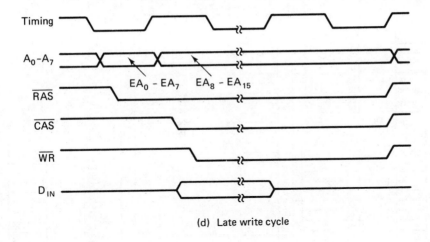

(d) Late write cycle

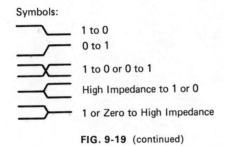

Symbols:

1 to 0

0 to 1

1 to 0 or 0 to 1

High Impedance to 1 or 0

1 or Zero to High Impedance

FIG. 9-19 (continued)

9-5 THE MEMORY SUBSYSTEM

Most fairly large computers have a memory subsystem that takes care of all memory storage. Figure 9-20 illustrates such a circuit that could be mounted on a printed circuit card with fingers that plug into a card edge connector. The connector is shown at the extreme left and extreme right of the figure. The memory card has inputs of a 16-bit address bus and an 8-bit data bus and outputs to an 8-bit data bus. Note that separated DATA IN and DATA OUT buses are used.

Line Receivers and Drivers

The purpose of the line receivers in the figure is to reject noise and accept the signal. These receivers are Schmitt triggers having considerable hysteresis. In addition, the line receiver should draw very little current from the bus so that many receivers on different cards can be connected in parallel to the same bus.

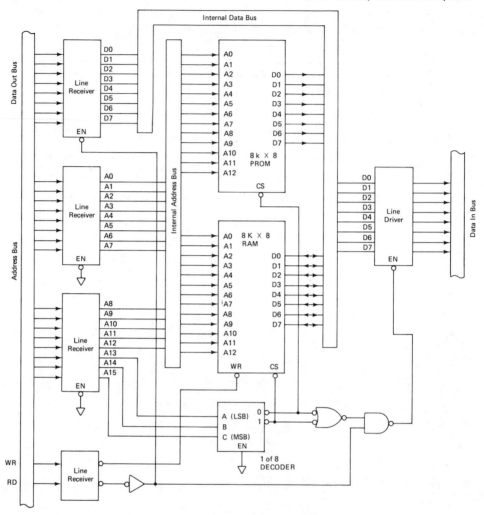

FIG. 9-20 RAM/PROM Memory Subsystem

Line drivers (see the right of the figure) should load the bus very little in their idle state. Thus, they are commonly tri-state devices going to a low impedance state only when told to do so. When on, the line driver must have sufficient current drive to overcome any loading or capacitance the bus may have.

Addressing the Card

In many computer systems there are multiple cards. Thus, each card must be designed so that it responds to its own particular address. The address selection on this particular card is done by the one of eight decoder. Its 0 output goes low when *CBA* is 000 and its 1 output goes low when *CBA* is 001. Since *CBA* represents the three most significant bits of the address bus, the 0 output responds to addresses from 0000 0000 0000 0000 to 0001 1111 1111 1111 or 0000 to 1FFF hex. In a similar manner, the 1 output responds to addresses from 0010 0000 0000 0000 to 0011 1111 1111 1111, or 2000 to 3FFF hex.

However, let us assume we must plug several memory cards into parallel address and data buses. We would want only one card at a time to respond. We might add the circuit of Fig. 9-21 in series with the A15 and A14 address lines. By setting the switches on each card to a different setting, we could have each card occupy a different memory space [Fig. 9-21(b)]. In this figure, the hexadecimal number represents the addresses to which the 0 output and the 1 output of the one of eight decoder will respond.

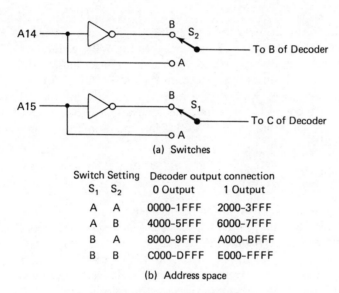

(a) Switches

Switch Setting		Decoder output connection	
S_1	S_2	0 Output	1 Output
A	A	0000–1FFF	2000–3FFF
A	B	4000–5FFF	6000–7FFF
B	A	8000–9FFF	A000–BFFF
B	B	C000–DFFF	E000–FFFF

(b) Address space

FIG. 9-21 Board Address Switches

The Memory

The memory of Fig. 9-21 consists of an 8k-byte PROM and an 8k-byte RAM. The chips have been wired such that they occupy contiguous memory addresses, the PROM occupying the lower half.

Reading

To read from the memory subsystem, an address is placed on the address bus and RD is raised. When this occurs, the line receiver and an inverter double invert RD and this high signal is fed to the DATA OUT bus line receiver, causing it to go into its high impedance state. Meanwhile, this high RD signal is fed to a **NAND** gate.

The other input of this gate is high if any address space on this card is being addressed. Thus, the output of the **NAND** gate goes low enabling the DATA IN line driver. This connects the internal data bus to the DATA IN bus.

Meanwhile, one of the outputs of the one of eight decoder is low enabling either the PROM or the RAM to be read. When $\overline{CS}$ is low, the chip will output data; when $\overline{CS}$ is high, the data output will go to high impedance. Thus, data from only one memory will go into the internal data bus and, from there, the external data bus. When the computer receives the data, it will lower RD, ending the read cycle. It usually waits a preset time, then assumes that the data is correct and strobes it into the processor portion of the computer, then lowers RD.

Writing

To write to this memory card, an address is placed on the address bus, the data to be written on the DATA OUT bus, and then the WR signal raised. Note that the DATA OUT line receiver is enabled when not reading. Thus, data appears on the internal data bus as soon as it is placed on the DATA OUT bus. In this manner, it precedes $\overline{WR}$, and there will not be a race between $\overline{WR}$ and the data to get to the memory chip. To complete the write cycle, the computer then lowers WR and, after that, removes its data from the DATA OUT bus and its address from the address bus.

9-6 SUMMARY

Each memory has an address bus and a data bus. Read only memories (ROMs) are meant to be read but not written whereas random-access memories (RAMs) can be both read and written. Dynamic RAMs must be refreshed periodically but consist of simpler circuits making them less expensive. Line receivers are meant to receive a signal from a bus, recovering the data from a relatively noisy environment. Line drivers are low output impedance devices used to drive a bus.

9-7 PROBLEMS

9-1. How many different address locations can be selected using (a) 6 address lines? (b) 16 address lines?

9-2. How many different address locations can be selected using (a) 12 address lines? (b) 32 address lines?

9-3. What is a bus? Is it parallel or serial?

9-4. In Fig. 9-5, if $E1$ is high, what happens to the data leads $D_0 - D_7$?

9-5. In Fig. 9-6, if D_O of U_1 is outputting a logical 1, what prevents D_O of U_2 from outputting a 0?

9-6. How does the gate of an ultraviolet (UV) EPROM obtain its charge?

9-7. Why does not the charge on a UV EPROM leak off?

9-8. Using a bipolar 32×8 PROM, implement the following equations simultaneously:

$$F_1 = \sum m(0, 1, 3, 5, 7, 21)$$
$$F_2 = \sum m(2, 3, 4, 7, 11, 12, 15, 26)$$
$$F_3 = \sum m(8, 9, 11, 13, 20, 23)$$

9-9. Use a 32×8 bipolar PROM to implement the equations:

$$F_1 = \sum m(2, 3, 6, 16, 26) + d(4, 5)$$
$$F_2 = \sum m(6, 7, 8, 10, 13, 14) + d(30, 31)$$
$$F_3 = \sum m(10, 11, 12, 13, 25) + d(0, 1, 3, 31)$$

9-10. Use a 32×8 bipolar PROM to translate 5-bit binary code into 5-bit Gray code.

9-11. Compare the number of transistors and resistors per cell required of a bipolar RAM cell, a MOS static RAM cell, and a MOS DRAM cell.

9-12. What is the major advantage of DRAM ICs over static RAMs? The disadvantage?

9-13. When does writing actually occur (a) in a DRAM early WRITE cycle? (b) In a DRAM late WRITE cycle?

9-14. What are the advantages and disadvantages of the multiplexing scheme used on DRAMs?

9-15. What are the two purposes of a line receiver?

9-16. What are the two purposes of a line driver?

9-17. We included address lines $A15$ and $A14$ in the decoding. Since it only takes one address line ($A13$) to differentiate between the PROM and the RAM, why were $A15$ and $A14$ examined by the logic of Fig. 9-21?

9-18. Assume the PROM responds to addresses 0000-1FFF. To what output of the one of eight decoder should the $\overline{CS}$ of the RAM be connected in order that it might respond to address 6000-7FFF (Fig. 9-21)?

9-19. Since reading only occurs when RD is true, and RD is used to turn on the DATA IN line driver, what prevents a logic race from occurring between the data transmission and the time at which RD goes high to low (Fig. 9-21)?

9-20. Why was the $\overline{EN}$ lead of the DATA OUT bus receiver connected to RD instead of $\overline{WR}$ (Fig. 9-21)?

THE DIGITAL COMPUTER

10

10-1 THE COMPUTER AGE

It has been said that we are living in a computer age, an era when, it seems, everything is being controlled by computers: our bank accounts, paychecks, travel reservations, credit card purchases, and even our income tax. Just what is this seemingly all-knowing machine? In this chapter we shall study the computer—what it is and just what makes it tick.

10-2 THE COMPUTER SYSTEM

As is true with most electronic systems, a computer system can easily be broken down into its various components (Fig. 10-1). The arithmetic and logic unit (ALU) performs all the necessary adding, subtracting, shifting, and logical operations. It requires one or two numbers upon which to operate and produces a result. For example, it could add 1011 (number A) to 0010 (number B) and obtain a result, 1101. The two numbers upon which it operates are called operands. In most computers, one operand is obtained from the memory and the second operand from the accumulator (this is a one-word memory) and the result placed back into the accumulator or memory.

The memory of a computer system contains two types of information: data and instructions. The operand discussed in the previous paragraph is an example of data. Instructions tell the computer what to do—add, subtract, multiply, divide, move data from accumulator to memory or memory to output or output to accumulator. These instructions, which are nothing more than binary numbers, pass from the memory to the control unit. Here, the control unit decodes each instruction, sending out signals

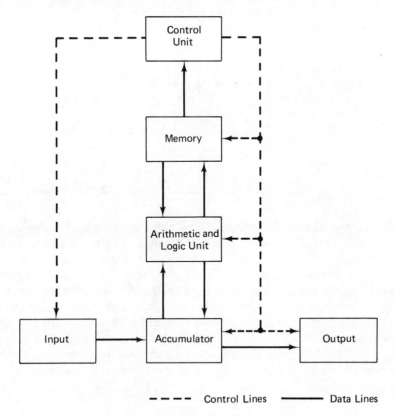

FIG. 10-1 Computer Organization

to all other system components, causing them to open gates, close gates, and so forth·
in order to *execute* the instruction, that is, to perform the operation specified by the
instruction.

To communicate with the computer system, input and output equipment is
required. These devices are also called peripherals. Input devices such as card readers,
paper tape readers, magnetic tape, and magnetic disks provide data for the system to
operate upon. Output devices such as card punches, video terminals, and line printers
allow the computer to communicate its answers to the human operator.

Imagine a series of pigeonholes such as are in Fig. 10-2. Assume each pigeonhole
represents a memory location within the computer. Thus, this memory contains 20
locations, or addresses. Next, remove the slip of paper from pigeonhole 1. It reads
"Take the number in location 15, add it to the number in location 16, and put the result
in location 10." Moving to location 15, we find the number 4 printed on its slip of
paper; location 16 contains a 5. Therefore, after replacing the slips, we add $4 + 5$,
giving 9, and write the number 9 on the slip of paper in location 10. Note that location 10 no longer contains the number it had before—it now contains a 9.

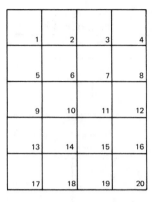

FIG. 10-2 Pigeonholes

Upon completion of this instruction, we move on to the next sequential instruction location, 2. It says "go to location 17." We therefore go to 17, remove its paper, and it says "Halt."

What we have done is to execute a computer program, a series of computer instructions. The memory was the set of pigeonholes. We had no input or output, because all the information originated within the system and stayed within the system. We acted as the control unit, for we read each instruction, interpreted it, and then performed the actual operation. Our operands for the first instruction were the numbers 4 and 5. Note that the pigeonholes contained two types of information: instructions and data. This is precisely what every computer system is: a machine for executing instructions.

10-3 COMPUTER INSTRUCTIONS

Since each computer system executes instructions, we cannot know the system without knowing the set of instructions. Note that each system has its own unique set. A set of instructions is shown in Table 10-1 for a fictitious computer we shall call the

TABLE 10-1. FIC-1 Instruction Set

Mnemonic	Word Format		Description
	Op Code	Address Field	
LOAD	0 0 1	A A A A A	Load the AC
STORE	0 1 0	A A A A A	Store the AC
ADD	0 1 1	A A A A A	Add to the AC
JUMP	1 0 0	A A A A A	Jump
JAZ	1 0 1	A A A A A	Jump if AC is zero
HALT	1 1 1	0 0 0 0 0	Halt
IAC	1 1 1	0 0 0 0 1	Increment AC
CAC	1 1 1	0 0 0 1 0	Complement AC

FIC-1. These instructions are typical of those in some systems. Note that each of the first five instructions is divided into a three-bit "op code" (operation code) field and a five-bit address field. These instructions all have to obtain data from memory, send data to memory, or find an address from memory. They are therefore called *memory reference* instructions. The last three instructions require no memory data and are, therefore, not memory reference instructions. The following details each instruction:

LOAD   001AAAAA   Load AC

   This instruction causes an eight-bit data word to be transferred from the memory location specified by AAAAA to the accumulator (AC). The data in the memory stays there; however, the previous contents of the AC are lost. As an example, 001 00101 would cause data to be transferred from location 5 of the memory (00101) to the AC.

STORE   010AAAAA   Store AC

   This instruction causes data to be written from the AC into the memory location specified by AAAAA. If the AC contained a 0000 1010 (a decimal 10) and the instruction 010 01111 were executed, the 10 of the AC would be written into location 15 (1111) of the memory. The AC remains unchanged, but the previous contents of memory location 15 are lost.

ADD   011AAAAA   Add AC

   This instruction will cause the AC to be added to the data contained in location AAAAA. The result will be put into the AC. The memory remains unchanged, but the AC's previous contents are lost.

JUMP   100AAAAA   Jump

   This instruction will cause the computer to start taking its instructions from location AAAAA. Memory and AC remain unchanged.

JAZ   101AAAAA   Jump if AC is 0

   This instruction will cause the computer to jump to location AAAAA and start taking its instructions from that location if the contents of the AC are 0. If the AC is not 0, the computer will take its next instruction from the location after the one containing the JAZ instruction.

HALT   11100000   Halt

   The computer will step executing instructions when it encounters a HALT instruction.

IAC   11100001   Increment the AC

This instruction will cause the number in the accumulator to be incremented (increased) by 1. If the AC contains a 1111 1111 and an IAC is encountered, it will go to 0000 0000.

CAC   11100010   Complement AC

This will cause the number in the accumulator to be complemented; that is, all its 1's will be changed to 0's and all its 0's to 1's.

We are now ready to write a small program for the FIC-1. Let us perform the operation

$$15 + 9 - 2 = ?$$

First, we shall put a 15 in memory location 10, the 9 in location 11, and the 2 in location 12 (Table 10-2). We shall start our program at memory location 0. The program will first load the data from memory location 12 into the AC (the AC will then equal 2_{10}). Next, the AC is complemented and then incremented. The FIC-1 contains no subtract, so we must find the two's complement and add. The AC now contains 1111 1110, a -2 in two's complement.

Next, the contents of locations 10 and 11 are added to the AC, resulting in 13_{10} and 22_{10}, respectively. Thus, our answer, 22, now is in the AC. We then store that answer in location 13 and halt. When done, the answer is both in memory and the AC.

TABLE 10-2. Programming the FIC-1

Location	Contents	Description	
0	001 01100	LOAD	12
1	111 00010	CAC	
2	111 00001	IAC	
3	011 01010	ADD	10
4	011 01011	ADD	11
5	010 01101	STORE	13
6	111 00000	HALT	
7			
8			
9			
10	000 01111		
11	000 01001		
12	000 00010		

Example 10-1: In certain applications, a computer must be programmed with a delay loop to prevent it from executing an instruction prior to some event occurring outside the computer. Design a delay loop for the FIC-1.

SOLUTION:

The INC, JUMP, and JAZ instructions will be required (Table 10-3). We shall start the program at location 2. Note that the program may start at any location. The number 0 is first loaded into the AC. The AC is then incremented. If the result is 0, the program will jump to location 6. If not, it will execute the JUMP 3 instruction and jump to location 3. This process continues until the AC is 1111 1111. At that time, adding 1 to the AC results in a 0000 0000, which will cause a jump to location 6 and the rest of the program.

TABLE 10-3. Delay Loop

Location	Contents	Description	
0			
1	000 00000		
2	001 00001	LOAD	1
3	111 00001	IAC	
4	101 00110	JAZ	6
5	100 00011	JUMP	3
6	– – – –	– –	

10-4 PROCESSING INSTRUCTIONS

Figure 10-3 is a detailed block diagram of the FIC-1 computer, along with abbreviations for each block.

The AC is a one-word register that *accumulates* the result of an operation. It may be incremented or complemented or loaded from the ALU on an ADD or LOAD instruction. Its information may be stored into MEM via the ALU and MD.

The ALU is a circuit composed of combinational logic that performs all arithmetic. The ADD is performed by obtaining one operand from the AC and the other from MD. Its result can then be transferred to the AC. On a LOAD or STORE instruction, it provides a straight-through data path between the AC and the MD.

The MA register contains the address of the memory location being read or written. When an instruction must be obtained, the MA is loaded from the PC. When data must be obtained, it is loaded from the IR.

The MD register contains the data being written or read from MEM. If it just obtained an instruction, the contents of MD are transferred to IR. If it just read a data word, the information is passed to the ALU.

The PC, IR, and RUN FF from the CONTROL section of the computer. The PC register contains the address of the next instruction to be processed. Except for a JUMP or JAZ (when the AC is 0), the PC is incremented after each instruction. If the computer is processing an ADD at location 13, PC contains a 14. A JUMP or JAZ (when the AC is 0) will cause the address portion of the instruction to be transferred from the IR to the PC, so that the next instruction will be taken from the jumped-to address.

The IR contains the instruction presently being processed. This register is always loaded from the MD. Control gates (not shown) then decode the IR, sending out signals to open gates, close gates, or cause a memory read or write.

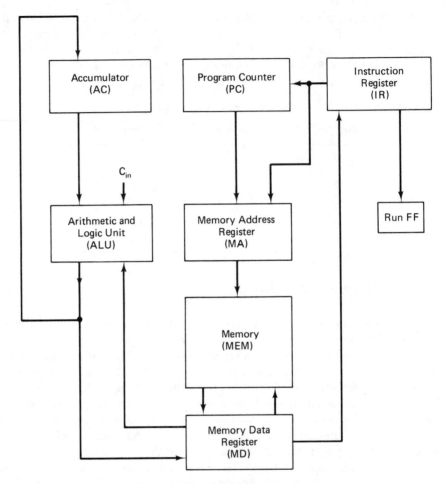

FIG. 10-3 FIC-1 Detailed Block Diagram

The RUN FF is set to a 0 when a HALT occurs, turning off the system clock. It is set to a 1 when the system is started (usually by a switch on the computer front panel).

The processing of each instruction can be divided into two *cycles:* a fetch cycle and an execute cycle. During the fetch cycle, the contents of the PC are transferred to the MA, the MA cause MEM to read a word from MEM to MD (the instruction itself), the MD is transferred to the IR, and the PC is incremented by 1 to get ready for the next instruction (Table 10-4).

The actual ADD, STORE, LOAD, etc., occur during the execute cycle. Since each type of instruction is unique, each will require that different operations be performed in order to execute the instruction. For example, to perform an ADD instruction,

TABLE 10-4. Instruction Processing

(a) FETCH Cycle:

PC → MA
MA → MEM
MEM → MD
MD → IR
PC + 1 → PC

(b) EXECUTE Cycle:

LOAD:	STORE:	ADD:
IR_{0-4} → MA	IR_{0-4} → MA	IR_{0-4} → MA
MA → MEM	AC → ALU	MA → MEM
MEM → MD	ALU → MD	MEM → MD
MD → ALU	MD → MEM	MD → ALU
ALU → AC		AC → ALU
		ALU → AC

JUMP:

IR_{0-4} → PC

JAZ:

If AC = 0, then IR_{0-4} → PC
If AC ≠ 0, then do nothing

HALT:

0 → RUN FF

IAC:

AC → ALU
C_{in} = 1 → ALU
ALU → AC

CAC:

AC → ALU
$\overline{ALU}$ → AC

A. The address field of the IR must be transferred to the MA.

B. The MEM must supply an operand to MD and from there to the ALU.

C. The AC must supply the second operand to the ALU.

D. The result must be transferred to the AC.

The execute cycle of all the instructions for the FIC-1 is shown in Table 10-4. Note that the fetch cycle is identical for all.

10-5 THE MC6801 MICROCOMPUTER

The Motorola MC6801 is a single chip computer consisting of a clock, processor, parallel input-output unit, timer, 128 bytes of RAM, and 1k bytes of ROM, all on one chip (Fig. 10-4). An 8-bit data bus and a 16-bit address bus are used internally to the chip and may be fed through to ports 3 and 4. An external crystal controls the onboard clock generating timing for the entire chip. The microprocessor unit (MPU) processes instructions it obtains from the ROM, storing and retrieving data from the RAM. The processor portion of the chip can talk to a parallel input-output (IO) section. In so doing it can send an 8-bit word to an external device or receive an 8-bit word. Each bit is sent or received simultaneously with the 7 other bits using eight different pins. Each port can receive or transmit 8 bits (port 2 is good for 5 bits).

The chip also contains a timer. Although it can be used in several modes, it can be thought of as a divide by n counter where n is programmable. Its input can be from an external source of clock pulses or from the processor clock and its output can be fed to the external world or to the processor, where it can be used much as an alarm clock, telling the processor to attend to some function at a predetermined time.

The serial communications interface (SCI) is a two way communications link that uses one wire to transmit and one to receive (ground is common). Information is received and transmitted in the form of pulse trains. By connecting these to a modem

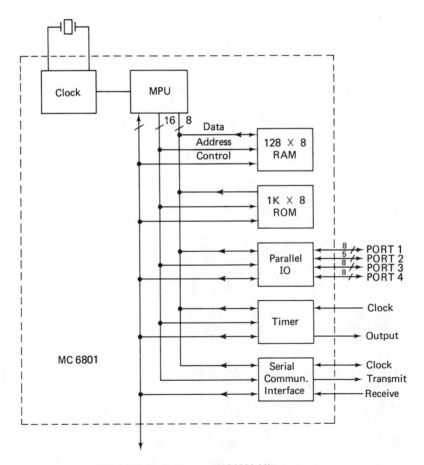

FIG. 10-4 Block Diagram MC6801 Microcomputer

(modulator-demodulator) and feeding this to a phone line, data can be transmitted and received over a phone line. Many printers and terminals require serial information. A clock can be inputted or outputted to the SCI to synchronize data.

Operating Modes

The MC6801 can operate in eight different modes. These modes can be used to add external ROM, external RAM, or to service IO devices. The modes can be classified into three fundamental types.

A. Single chip.
B. Expanded non-multiplexed.
C. Expanded multiplexed.

The single chip mode (Fig. 10-5) configures the chip to 8-bit IO ports and a 5-bit IO port that can include serial data and the timer. This mode is used when all of memory

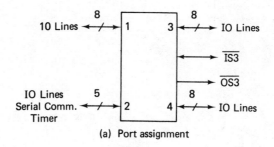

(a) Port assignment

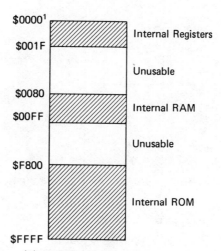

[1]The $ preceding a number indicates that number is hexadecimal.

FIG. 10-5 Single Chip Mode

(b) Mode 7 addressing space

is inside the MC6801 chip. Mode 7 is typical of this configuration and assigns address space, as shown in Fig. 10-5(b).

The expanded non-multiplexed mode permits addressing of up to 256 more bytes of RAM or ROM external to the MC6801. Figure 10-6(a) illustrates the bus structure. Port 4 forms an 8-bit address bus and port 3 forms an 8-bit data bus. Signal R/$\overline{\text{W}}$ selects either a READ or WRITE cycle for the external memory and $\overline{\text{IOS}}$ goes true when this memory is to be accessed. Thus, R/$\overline{\text{W}}$ can be connected to the R/$\overline{\text{W}}$ input of external memory and $\overline{\text{IOS}}$ can be connected to $\overline{\text{CS}}$ of the external memory. Mode 5 is used and assigns address space, as shown in Fig. 10-6(b).

The expanded multiplexed modes (0, 1, 2, 3, and 6), permit a 16-bit address bus and an 8-bit data bus to access external memory [Fig. 10-7(a)]. Ports 3 and 4 first send

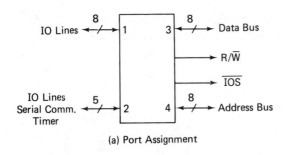

(a) Port Assignment

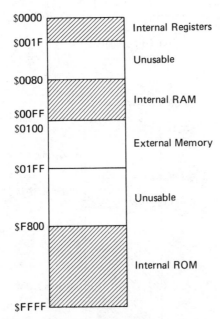

(b) Mode 5 Addressing Space   **FIG. 10-6** Expanded, Non-Multiplexed Mode

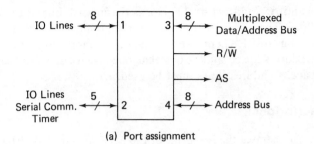

(a) Port assignment

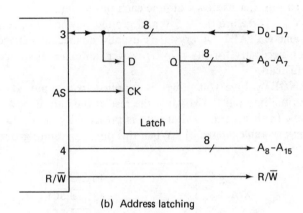

(b) Address latching

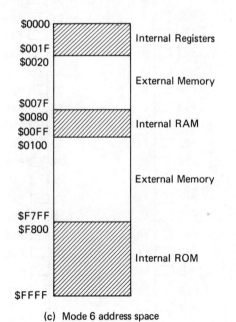

(c) Mode 6 address space

FIG. 10-7 Expanded Multiplexed Mode

out an address and AS (address strobe) clocks the lower 8 bits of the address into a latch [Fig. 10-7(b)]. The lower order address is then removed from port 3 and these 8 bits act as a data bus to READ or WRITE data to memory. Figure 10-7(c) illustrates one of several address space allocations that may be used. Note that, in order to address this space, 16 address lines must be used.

Pin Descriptions

Figure 10-8 illustrates the footprint of the MC6801 and Fig. 10-9 illustrates the function of each pin. Let us now examine each pin in detail.

V_{cc} **and** V_{ss} These are the $+5$ V and ground power leads, respectively.

XTAL1 and EXTAL2   A crystal is usually connected between these pins to create the clock necessary for the chip. However, an external source of clock pulses may be used, instead.

V_{cc} **STANDBY**   Five volt power is supplied to this pin. It can supply keep-alive current to the internal RAM while the rest of the chip is powered down (V_{cc} is at ground level). In this manner, vital data is preserved with a minimum of current.

$\overline{\text{NMI}}$ **(non-maskable interrupt)**   When this pin is driven to ground, the processor

Vss	1	40	E

```
        Vss  ⊏  1        40 ⊐  E
       XTAL1 ⊏  2        39 ⊐  SC1
      EXTAL 2 ⊏ 3        38 ⊐  SC2
        NMI  ⊏  4        37 ⊐  P30
        IRQ1 ⊏  5        36 ⊐  P31
       RESET ⊏  6        35 ⊐  P32
        Vcc  ⊏  7        34 ⊐  P33
        P20  ⊏  8        33 ⊐  P34
        P21  ⊏  9        32 ⊐  P35
        P22  ⊏ 10        31 ⊐  P36
        P23  ⊏ 11        30 ⊐  P37
        P24  ⊏ 12        29 ⊐  P40
        P10  ⊏ 13        28 ⊐  P41
        P11  ⊏ 14        27 ⊐  P42
        P12  ⊏ 15        26 ⊐  P43
        P13  ⊏ 16        25 ⊐  P44
        P14  ⊏ 17        24 ⊐  P45
        P15  ⊏ 18        23 ⊐  P46
        P16  ⊏ 19        22 ⊐  P47
        P17  ⊏ 20        21 ⊐  Vcc
                                  Standby
```

FIG. 10-8 MC6801 Pin Assignment (Courtesy Motorola Semiconductor Products, Inc.)

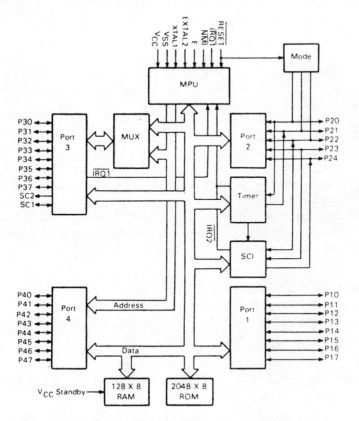

FIG. 10-9 MC6801 Microcomputer Block Diagram (Courtesy Motorola Semiconductor Products, Inc.)

will stop what it is doing, get the two bytes at addresses $FFFC and $FFFD, and use these as an address. The processor will then execute the instruction found at this new address. Meanwhile, the processor remembers at what point it was interrupted and can return to that point by executing an RTI (return from interrupt) instruction. Interrupts are used to service an external device upon demand by that external device.

IRQ1 (interrupt request 1) This lead is also an interrupt pin. However, the processor will ignore this interrupt if it previously executed an SEI (set interrupt mask) instruction. The processor permits itself to be interrupted by executing a CLI (clear interrupt mask) instruction. Should the mask be cleared and IRQ1 be brought low, the processor will obtain 2 bytes from $FFF8 and $FFF9, and use this 16-bit number as an address. The processor will then execute the instruction found at this new address. It can return to the point at which it was interrupted by executing an RTI (return from interrupt) instruction.

RESET This input resets the MC6801.

E (enable) This pin contains an output clock that is fed to other MC6801 compatible chips to synchronize them with the MC6801.

SC1 and SC2 (strobe control) These are two control leads, the use of which depends upon the mode selected. The two control leads shown in each of Figs. 10-5, 10-6, and 10-7 are actually SC1 and SC2.

P10-P17   These eight bits form port 1, bits 0 through 7. The designation "P" indicates a port, the left digit is the port (1, 2, or 3), and the right digit is the bit of that particular port. Thus, P13 represents bit 3 of port 1. These ports can be configured as inputs or outputs by the program.

P20-24   These five pins form port 2. The MPU strobes the logic level of P20-P22 into itself at the trailing edge of $\overline{\text{RESET}}$. These three bits determine the mode of the MC6801; however, after $\overline{\text{RESET}}$, they become IO pins. Note that, according to Fig. 10-7, P23 can be programmed to be the input to the serial communications interface (SCI), and P24 as the SCI output. P22 can be programmed to be an input clock or an output clock for the SCI. P21 can be used as an output from the timer, thus providing pulses at a programmed rate to an external device. P20 may be used as the source of input pulses to the timer. When these options are not programmed, P20-P24 can be used as a five-bit input or output port.

P30-P37   These eight bits form port 3 and may be used as:
A. Input or output pins (single chip mode),
B. Data bus (expanded non-multiplexed mode) or,
C. Multiplexed data/address bus (expanded multiplexed mode),

P40-47   These eight leads may be used as:

A. Input or output pins (single chip mode) or,
B. Address output (expanded non-multiplexed and expanded multiplexed modes)

10-6 THE MC6801 INSTRUCTION SET

The MC6801 has a rich variety of instructions within its set allowing the programmer to easily form his or her paragraphs from the processor vocabulary. These instructions depend very heavily upon the registers provided within the processor.

Programming Model

Figure 10-10 illustrates the various registers to which the programmer has access. There are two eight-bit accumulators, A and B. In general, instructions take the data within A or B, combine it with data in memory, and place it back into A or B. Some instructions treat A and B as one single 16-bit accumulator called D.

The model also has a 16-bit index register, X. This register usually contains a memory address. It may be incremented or decremented, allowing access to an entire table of sequential data within the memory.

The SP (stack pointer) is a 16-bit register containing the address of a RAM location called the stack. This stack is much like a stack of plates used in a restaurant.

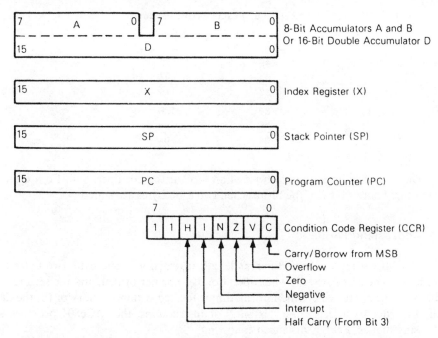

FIG. 10-10 MC6801 Programming Model (Courtesy Motorola Semiconductor Products, Inc.)

We may add to the stack by placing more plates on the stack; however, the first one we take off the stack is the last one we put on the stack. The SP register used within the MC6801 is decremented every time data is placed on the stack and incremented every time data is removed from the stack. Instructions permit the programmer to push data onto the stack (PHS or PSHX) or pull it off the stack (PUL and PULX). The stack is also used much as the crumbs of Hansel and Gretel to store the return address from a subroutine or from servicing the two interrupt inputs.

The PC (program counter) contains the address of the next instruction to be executed. Usually, it increments one at a time through the program. However, jump instructions permit loading any number into the PC, transferring control to any part of memory.

The CCR (condition code register) contains eight bits. The two most significant bits are always 1's. Bit 5, H, is the half carry. If two bytes of data are added and there is a carry from bit 3 to 4 of the result, H is set to a 1. If there is no half carry, H is set to 0. The I bit is set to a 1 by SEI (set interrupt mask), the result of which permits the processor to ignore interrupts on the $\overline{\text{IRQI}}$ lead. The I bit is cleared by a CLI instruction, permitting $\overline{\text{IRQI}}$ interrupts to be recognized. The N (negative) bit of the CCR is set to a 1 if the result of the previous arithmetic operation is negative and is set to 0 if the result is positive. The V (overflow) bit is set to a 1 if the sum of two positive numbers was negative or the sum of two negative numbers was positive. Table 10-5 illustrates these two cases.

TABLE 10-5. Examples of Overflow

Hexadecimal	Decimal
4F	+79
43	+67
92	−110
80	−128
84	−124
04	+04

The C (carry) bit is the ninth bit of an 8-bit arithmetic operation. For example, it is the carry out of bit 7 of the result when two 8-bit bytes are added.

Addressing Modes

The MC6801 uses three different length instructions: one byte, two bytes and three bytes. The first byte tells what operation is to be performed, and the second and third bytes specify (a) where the information is located within memory or (b) the data itself. In order to increase the flexibility of instructions, the MC6801 provides six addressing modes. Each is discussed following.

Immediate Addressing   This is an addressing mode using two-or three-byte instructions in which the first byte specifies the operation to be performed and the second (and third) bytes specify the data to be operated upon. For example, in the ADDA (add to register A) immediate instruction, the first byte, $8B, specifies an ADDA immediate instruction and the second byte specifies what data is to be added. In executing, this instruction takes the byte in the A register, adds it to the second byte of the instruction, and places the result back into A. It can be expressed as:

$$A + B2 \longrightarrow A$$

An LDX (load index register) immediate instruction contains three bytes. The first is the operation (op) code; the second and third byte form the 16-bit operand that is to be placed into the X register. We can express this as:

$$B2, B3 \longrightarrow X$$

Direct Addressing   This mode uses two-byte instructions in which the first byte is the op code and the second byte specifies an 8-bit address at which one of the operands is located. The upper 8 bits of the address are considered all zero. A SUBA (subtract A) instruction, for example, has the op code $90 for its first byte and a memory address for its second byte:

$$A - (B2) \longrightarrow A$$

The parentheses around the $B2$ mean that $B2$ is not the data but the address of the data. If this second byte were $3F, the location of the data that is to be subtracted from A would be $003F. The result would be placed in A.

Extended Addressing   This mode uses three-byte instructions in which the first byte is the op code, and the second and third bytes form a 16-bit address field. For example, a CLR (clear) instruction in the extended mode has $7F for its first byte followed by two bytes of address. These two bytes tell the processor the location that is to be cleared to zero. If the second and third bytes were $20 and $46 respectively, location $2046 would be cleared to zero when this instruction is executed. The operation can be expressed as:

$$\$00 \longrightarrow (B2, B3)$$

Indexed Addressing   The plot thickens. This mode uses two-byte instructions in which the first byte contains the op code and the second byte contains an unsigned integer called an offset. To locate the operand, this offset is added to the index register to form the effective address:

$$X + B2 \longrightarrow EA$$

The operand is located at this effective address. Let us try an ADDB (add to B) indexed instruction. We shall assume the second byte of the instruction is a $13 and the index register contains a $3241. The instruction would be stored as:

First byte:	$EB	This is the op code.
Second byte:	$13	This is the offset.

Upon execution, the processor would locate the data and perform the operation according to the equation:

$$B + (X + B2) \longrightarrow B$$
$$B + (\$3241 + \$0013) \longrightarrow B$$
$$B + (\$3254) \longrightarrow B$$

The processor would add the number at location $3254 to B and place the result in the B register.

Inherent Addressing   This mode consists of single byte instructions that do not need to specify an address within memory. Examples include:

ASRA	Arithmetic shift right by one bit the A accumulator
INCB	Increment the B accumulator by one
SEC	Set the carry bit

Relative Addressing   The relative addressing mode is used for many of the jump instructions. These consist of two bytes: an op code and an offset. The offset is considered a signed integer and is added to the program counter to obtain the address to which the program will jump. Let us consider a BCS (branch on carry set) instruction having the following two bytes at the locations indicated:

Location	Contents	Description
$3043	$25	Op Code
$3044	$F3	Offset
$3045	—	The PC would point to this location

If the carry bit is a 1, the $F3 is added to the program counter to form:

$$\$3045 + \$FFF3 = \$3038$$

The FF's were prefixed to the F3 to keep the second number negative. If we were to have added $00F3 instead of $FFF3, we would have been considering the operand as a positive integer instead of a signed integer. However, the processor will perform a jump to location $3038 in this example.

Instructions

Tables 10-6, 10-7, 10-8, and 10-9 illustrate the instruction set of the MC6801. Note that there are no INPUT or OUTPUT instructions. All ports are treated as memory locations and have pre-assigned addresses.

One Final Word

This discussion of the MC6801 has been, of necessity, abbreviated. For more precise details consult the manuals and specification sheets provided by the manufacturer.

10-7 PERIPHERAL EQUIPMENT

Computer systems must be able to talk to people and receive input form people. This is the purpose of input/output equipment. In addition, since the RAM cannot possibly store an infinite amount of information, mass storage devices must be used to reduce the cost per bit of storage. These two types of equipment are referred to as peripherals and will be discussed in this section.

Card Systems

Punched card systems have the advantage of easily editing input data. A variety of card processing equipment is also available: key punches for data entry, sorters for putting the cards in order, interpreters for printing alphanumeric characters along the top edge of the card, and readers and punches which control and are controlled by the computer.These card systems use the Hollerith code described in Chapter 3.

Card readers are usually photoelectric devices reading up to 1000 cards per minute. The stack to be read is first placed in the read *hopper*, and then one card at a time is read. Two methods are used to separate a single card from the read stack for reading. One uses a feed knife which is just thick enough so that it will remove one card. In the second method, a revolving belt presses against the bottom of the deck, removing a single card at a time.

The cards are then read by the photoelectric reader, a hole being a 1 and no hole

TABLE 10-6. MC6801 Index Register and Stack Manipulation Instructions (Courtesy Motorola Semiconductor Products, Inc.)

Pointer Operations	Mnemonic	Immed OP	~	#	Direct OP	~	#	Index OP	~	#	Extnd OP	~	#	Inher OP	~	#	Boolean/Arithmetic Operation	H (5)	I (4)	N (3)	Z (2)	V (1)	C (0)
Compare Index Reg	CPX	8C	4	3	9C	5	2	AC	6	2	BC	6	3				$X - M : M+1$	—	—	↕	↕	↕	↕
Decrement Index Reg	DEX													09	3	1	$X - 1 \rightarrow X$	●	●	↕	↕	●	●
Decrement Stack Pntr	DES													34	3	1	$SP - 1 \rightarrow SP$	●	●	●	●	●	●
Increment Index Reg	INX													08	3	1	$X + 1 \rightarrow X$	●	●	↕	↕	●	●
Increment Stack Pntr	INS													31	3	1	$SP + 1 \rightarrow SP$	●	●	●	●	●	●
Load Index Reg	LDX	CE	3	3	DE	4	2	EE	5	2	FE	5	3				$M \rightarrow X_H, (M+1) \rightarrow X_L$	●	●	↕	↕	R	●
Load Stack Pntr	LDS	8E	3	3	9E	4	2	AE	5	2	BE	5	3				$M \rightarrow SP_H, (M+1) \rightarrow SP_L$	●	●	↕	↕	R	●
Store Index Reg	STX				DF	4	2	EF	5	2	FF	5	3				$X_H \rightarrow M, X_L \rightarrow (M+1)$	●	●	↕	↕	R	●
Store Stack Pntr	STS				9F	4	2	AF	5	2	BF	5	3				$SP_H \rightarrow M, SP_L \rightarrow (M+1)$	●	●	↕	↕	R	●
Index Reg → Stack Pntr	TXS													35	3	1	$X - 1 \rightarrow SP$	●	●	●	●	●	●
Stack Pntr → Index Reg	TSX													30	3	1	$SP + 1 \rightarrow X$	●	●	●	●	●	●
Add	ABX													3A	3	1	$B + X \rightarrow X$	●	●	●	●	●	●
Push Data	PSHX													3C	4	1	$X_L \rightarrow M_{SP}, SP - 1 \rightarrow SP$ $X_H \rightarrow M_{SP}, SP - 1 \rightarrow SP$	●	●	●	●	●	●
Pull Data	PULX													38	5	1	$SP + 1 \rightarrow SP, M_{SP} \rightarrow X_H$ $SP + 1 \rightarrow SP, M_{SP} \rightarrow X_L$	●	●	●	●	●	●

See Condition Code Register Notes listed after Table 10-9

TABLE 10-7. MC6801 Accumulator and Memory Instructions (Courtesy Motorola Semiconductor Products, Inc.)

Accumulator and Memory Operations	MNE	Immed Op	~	#	Direct Op	~	#	Index Op	~	#	Extend Op	~	#	Inher Op	~	#	Boolean Expression	H	I	N	Z	V	C
Add Acmltrs	ABA													1B	2	1	A + B → A	•	—	↕	↕	↕	↕
Add B to X	ABX													3A	3	1	00:B + X → X	—	—	—	—	—	—
Add with Carry	ADCA	89	2	2	99	3	2	A9	4	2	B9	4	3				A + M + C → A	•	—	↕	↕	↕	↕
	ADCB	C9	2	2	D9	3	2	E9	4	2	F9	4	3				B + M + C → B	•	—	↕	↕	↕	↕
Add	ADDA	8B	2	2	9B	3	2	AB	4	2	BB	4	3				A + M → A	•	—	↕	↕	↕	↕
	ADDB	CB	2	2	DB	3	2	EB	4	2	FB	4	3				B + M → A	•	—	↕	↕	↕	↕
Add Double	ADDD	C3	4	3	D3	5	2	E3	6	2	F3	6	3				D + M:M + 1 → D	—	—	↕	↕	↕	↕
And	ANDA	84	2	2	94	3	2	A4	4	2	B4	4	3				A · M → A	—	—	↕	↕	R	—
	ANDB	C4	2	2	D4	3	2	E4	4	2	F4	4	3				B · M → B	—	—	↕	↕	R	—
Shift Left, Arithmetic	ASL							68	6	2	78	6	3					—	—	↕	↕	↕	↕
	ASLA													48	2	1		—	—	↕	↕	↕	↕
	ASLB													58	2	1		—	—	↕	↕	↕	↕
Shift Left Dbl	ASLD													05	3	1		—	—	↕	↕	↕	↕
Shift Right, Arithmetic	ASR							67	6	2	77	6	3					—	—	↕	↕	↕	↕
	ASRA													47	2	1		—	—	↕	↕	↕	↕
	ASRB													57	2	1		—	—	↕	↕	↕	↕
Bit Test	BITA	85	2	2	95	3	2	A5	4	2	B5	4	3				A · M	—	—	↕	↕	R	—
	BITB	C5	2	2	D5	3	2	E5	4	2	F5	4	3				B · M	—	—	↕	↕	R	—
Compare Acmltrs	CBA													11	2	1	A - B	—	—	↕	↕	↕	↕

320

Operation	Mnemonic	Immed OP	Immed ~	Immed #	Direct OP	Direct ~	Direct #	Index OP	Index ~	Index #	Extend OP	Extend ~	Extend #	Inher OP	Inher ~	Inher #	Boolean/Arithmetic Operation	H	I	N	Z	V	C
Clear	CLR							6F	6	2	7F	6	3				00 → M	•	•	R	S	R	R
	CLRA													4F	2	1	00 → A	•	•	R	S	R	R
	CLRB													5F	2	1	00 → B	•	•	R	S	R	R
Compare	CMPA	81	2	2	91	3	2	A1	4	2	B1	4	3				A - M	•	•	↕	↕	↕	↕
	CMPB	C1	2	2	D1	3	2	E1	4	2	F1	4	3				B - M	•	•	↕	↕	↕	↕
1's Complement	COM							63	6	2	73	6	3				$\bar{M}$ → M	•	•	↕	↕	R	S
	COMA													43	2	1	$\bar{A}$ → A	•	•	↕	↕	R	S
	COMB													53	2	1	$\bar{B}$ → B	•	•	↕	↕	R	S
Decimal Adj. A	DAA													19	2	1	Adj binary sum to BCD	•	•	↕	↕	↕	↕
Decrement	DEC							6A	6	2	7A	6	3				M - 1 → M	•	•	↕	↕	↕	•
	DECA													4A	2	1	A - 1 → A	•	•	↕	↕	↕	•
	DECB													5A	2	1	B - 1 → B	•	•	↕	↕	↕	•
Exclusive OR	EORA	88	2	2	98	3	2	A8	4	2	B8	4	3				A $\oplus$ M → A	•	•	↕	↕	R	•
	EORB	C8	2	2	D8	3	2	E8	4	2	F8	4	3				B $\oplus$ M → B	•	•	↕	↕	R	•
Increment	INC							6C	6	2	7C	6	3				M + 1 → M	•	•	↕	↕	↕	•
	INCA													4C	2	1	A + 1 → A	•	•	↕	↕	↕	•
	INCB													5C	2	1	B + 1 → B	•	•	↕	↕	↕	•
Load Acmltrs	LDAA	86	2	2	96	3	2	A6	4	2	B6	4	3				M → A	•	•	↕	↕	R	•
	LDAB	C6	2	2	D6	3	2	E6	4	2	F6	4	3				M → B	•	•	↕	↕	R	•
Load Double	LDD	CC	3	3	DC	4	2	EC	5	2	FC	5	3				M:M + 1 → D	•	•	↕	↕	R	•
Logical Shift, Left	LSL							68	6	2	78	6	3					•	•	↕	↕	↕	↕
	LSLA													48	2	1		•	•	↕	↕	↕	↕
	LSLB													58	2	1		•	•	↕	↕	↕	↕
	LSLD													05	3	1		•	•	↕	↕	↕	↕
Shift Right, Logical	LSR							64	6	2	74	6	3					•	•	R	↕	↕	↕
	LSRA													44	2	1		•	•	R	↕	↕	↕
	LSRB													54	2	1		•	•	R	↕	↕	↕
	LSRD													04	3	1		•	•	R	↕	↕	↕
Multiply	MUL													3D	10	1	A X B → D	•	•	•	↕	•	↕

See Condition Code Register Notes listed after Table 10-9

TABLE 10-7. (continued)

Accumulator and Memory Operations	MNE	Immed Op	~	#	Direct Op	~	#	Index Op	~	#	Extend Op	~	#	Inher Op	~	#	Boolean Expression	H	I	N	Z	V	C
2's Complement (Negate)	NEG							60	6	2	70	6	3				00 - M → M	●	●	↕	↕	↕	↕
	NEGA													40	2	1	00 - A → A	●	●	↕	↕	↕	↕
	NEGB													50	2	1	00 - B → B	●	●	↕	↕	↕	↕
No Operation	NOP													01	2	1	PC + 1 → PC	●	●	●	●	●	●
Inclusive OR	ORAA	8A	2	2	9A	3	2	AA	4	2	BA	4	3				A + M → A	●	●	↕	↕	R	●
	ORAB	CA	2	2	DA	3	2	EA	4	2	FA	4	3				B + M → B	●	●	↕	↕	R	●
Push Data	PSHA													36	3	1	A → Stack	●	●	●	●	●	●
	PSHB													37	3	1	B → Stack	●	●	●	●	●	●
Pull Data	PULA													32	4	1	Stack → A	●	●	●	●	●	●
	PULB													33	4	1	Stack → B	●	●	●	●	●	●
Rotate Left	ROL							69	6	2	79	6	3					●	●	↕	↕	↕	↕
	ROLA													49	2	1		●	●	↕	↕	↕	↕
	ROLB													59	2	1		●	●	↕	↕	↕	↕
Rotate Right	ROR							66	6	2	76	6	3					●	●	↕	↕	↕	↕
	RORA													46	2	1		●	●	↕	↕	↕	↕
	RORB													56	2	1		●	●	↕	↕	↕	↕
Subtract Acmltr	SBA													10	2	1	A - B → A	●	●	↕	↕	↕	↕
Subtract with Carry	SBCA	82	2	2	92	3	2	A2	4	2	B2	4	3				A - M - C → A	●	●	↕	↕	↕	↕
	SBCB	C2	2	2	D2	3	2	E2	4	2	F2	4	3				B - M - C → B	●	●	↕	↕	↕	↕
Store Acmltrs	STAA				97	3	2	A7	4	2	B7	4	3				A → M	●	●	↕	↕	R	●
	STAB				D7	3	2	E7	4	2	F7	4	3				B → M	●	●	↕	↕	R	●
	STD				DD	4	2	ED	5	2	FD	5	3				D → M:M + 1	●	●	↕	↕	R	●
Subtract	SUBA	80	2	2	90	3	2	A0	4	2	B0	4	3				A - M → A	●	●	↕	↕	↕	↕
	SUBB	C0	2	2	D0	3	2	E0	4	2	F0	4	3				B - M → B	●	●	↕	↕	↕	↕
Subtract Double	SUBD	83	4	3	93	5	2	A3	6	2	B3	6	3				D - M:M + 1 → D	●	●	↕	↕	↕	↕
Transfer Acmltr	TAB													16	2	1	A → B	●	●	↕	↕	R	●
	TBA													17	2	1	B → A	●	●	↕	↕	R	●
Test, Zero or Minus	TST							6D	6	2	7D	6	3				M - 00	●	●	↕	↕	R	R
	TSTA													4D	2	1	A - 00	●	●	↕	↕	R	R
	TSTB													5D	2	1	B - 00	●	●	↕	↕	R	R

TABLE 10-8. MC6801 Jump and Branch Instructions (Courtesy Motorola Semiconductor Products, Inc.)

Operations	Mnemonic	Direct OP	~	#	Relative OP	~	#	Index OP	~	#	Extnd OP	~	#	Inherent OP	~	#	Branch Test	H	I	N	Z	V	C
Branch Always	BRA				20	3	2										None	•	•	•	•	•	•
Branch Never	BRN				21	3	2										None	•	•	•	•	•	•
Branch If Carry Clear	BCC				24	3	2										$C = 0$	•	•	•	•	•	•
Branch If Carry Set	BCS				25	3	2										$C = 1$	•	•	•	•	•	•
Branch If = Zero	BEQ				27	3	2										$Z = 1$	•	•	•	•	•	•
Branch If ≥ Zero	BGE				2C	3	2										$N \oplus V = 0$	•	•	•	•	•	•
Branch If > Zero	BGT				2E	3	2										$Z + (N \oplus V) = 0$	•	•	•	•	•	•
Branch If Higher	BHI				22	3	2										$C + Z = 0$	•	•	•	•	•	•
Branch If Higher or Same	BHS				24	3	2										$C = 0$	•	•	•	•	•	•
Branch If ≤ Zero	BLE				2F	3	2										$Z + (N \oplus V) = 1$	•	•	•	•	•	•
Branch If Carry Set	BLO				25	3	2										$C = 1$	•	•	•	•	•	•
Branch If Lower Or Same	BLS				23	3	2										$C + Z = 1$	•	•	•	•	•	•
Branch If < Zero	BLT				2D	3	2										$N \oplus V = 1$	•	•	•	•	•	•
Branch If Minus	BMI				2B	3	2										$N = 1$	•	•	•	•	•	•
Branch If Not Equal Zero	BNE				26	3	2										$Z = 0$	•	•	•	•	•	•
Branch If Overflow Clear	BVC				28	3	2										$V = 0$	•	•	•	•	•	•
Branch If Overflow Set	BVS				29	3	2										$V = 1$	•	•	•	•	•	•
Branch If Plus	BPL				2A	3	2										$N = 0$	•	•	•	•	•	•
Branch To Subroutine	BSR				8D	6	2										} Special Operations	•	•	•	•	•	•
Jump	JMP							6E	3	2	7E	3	3					•	•	•	•	•	•
Jump To Subroutine	JSR	9D	5	2				AD	6	2	BD	6	3					•	•	•	•	•	•
No Operation	NOP													01	2	1		•	•	•	•	•	•
Return From Interrupt	RTI													3B	10	1		↕	↕	↕	↕	↕	↕
Return From Subroutine	RTS													39	5	1	} Special Operations	•	•	•	•	•	•
Software Interrupt	SWI													3F	12	1		•	S	•	•	•	•
Wait For Interrupt	WAI													3E	9	1		•	S	•	•	•	•

Condition Code Register columns: 5 = H, 4 = I, 3 = N, 2 = Z, 1 = V, 0 = C

See Condition Code Register Notes listed after Table 10-9

TABLE 10-9. MC6801 Condition Code Register Manipulation Instructions (Courtesy Motorola Semiconductor Products, Inc.)

Operations	Mnemonic	Inherent OP	~	#	Boolean Operation	5 H	4 I	3 N	2 Z	1 V	0 C
Clear Carry	CLC	0C	2	1	0 → C	●	●	●	●	●	R
Clear Interrupt Mask	CLI	0E	2	1	0 → I	●	R	●	●	●	●
Clear Overflow	CLV	0A	2	1	0 → V	●	●	●	●	R	●
Set Carry	SEC	0D	2	1	1 → C	●	●	●	●	●	S
Set Interrupt Mask	SEI	0F	2	1	1 → I	●	S	●	●	●	●
Set Overflow	SEV	0B	2	1	1 → V	●	●	●	●	S	●
Accumulator A → CCR	TAP	06	2	1	A → CCR	→	→	→	→	→	→
CCR → Accumulator A	TPA	07	2	1	CCR → A	●	●	●	●	●	●

LEGEND

OP Operation Code (Hexadecimal)
~ Number of MPU Cycles
MSP Contents of memory location pointed to by Stack Pointer

\# Number of Program Bytes
+ Arithmetic Plus
− Arithmetic Minus
● Boolean AND
X Arithmetic Multiply
+ Boolean Inclusive OR
⊕ Boolean Exclusive OR
$\overline{\text{M}}$ Complement of M
→ Transfer Into
0 Bit = Zero
00 Byte = Zero

CONDITION CODE SYMBOLS

H Half-carry from bit 3
I Interrupt mask
N Negative (sign bit)
Z Zero (byte)
V Overflow, 2's complement
C Carry/Borrow from MSB
R Reset Always
S Set Always
↕ Set Always
↨ Affected
● Not Affected

324

being a 0. From there, the card passes to the stacker, where it is added to the bottom of the deck of cards that have been read.

In some systems, the computer may output to a card punch capable of punching up to 300 cards per minute. Cards are punched a row at a time with the 12 row being punched first. Note that this requires that the electronics control 80 punch mechanisms at a time.

Line Printers

The line printer is the output workhorse on most large data processing systems; some units are capable of printing 1000 lines per minute. Line printers receive 132 characters of eight-bit parallel EBCDIC code (see Chapter 3) before printing. Some printers use ASCII code. The printing drum of the unit contains the complete character set in raised letters in each of the 132 columns (Fig. 10-11), each column having its own hammer and solenoid. The drum revolves at a constant speed, and when the correct letter is under the print hammer the solenoid energizes, causing the hammer to hit the paper, printing a character.

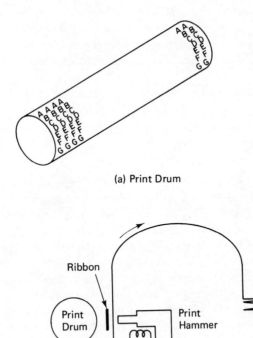

(a) Print Drum

Ribbon

Print
Drum

Print
Hammer

Paper
Supply

FIG. 10-11 Line Printer   (b) Hammer Mechanism

Some slower line printers cut cost by reducing the number of hammers and placing them on a moving mechanism that selects the proper column. This requires that the print drum revolve several times per line printed, reducing the number of lines per minute that can be printed.

Other Printers

The dot matrix impact printer forms a character out of 9 or more vertical dots. Each dot is printed by a solenoid controlled wire impacting the ribbon. Some units print only as the head travels left to right, whereas others print in both directions.

The electrostatic printer forms characters on a specially coated paper from dots created by a spark from the head to the paper.

The daisy wheel printer forms characters from a hammer striking a "petal" on the daisy wheel. Each petal has a separate character on it and the printer rotates the daisy to select the correct petal.

CRT Terminals

Cathode ray tube (CRT) terminals are very popular for remote entry and retrieval. Some examples include airline ticket reservations, student registration, and problem-solving terminals for students and engineers. Some law enforcement vehicles have CRT terminals that enable them to check whether a car is stolen or a suspect is wanted.

The CRT display has several advantages over a teletypewriter:

A. It makes very little noise.
B. It requires no paper or ribbon changes.
C. It "prints" faster.
D. It is fully electronic.

It does, of course, have the disadvantage of providing no printout (although some CRT terminals provide a photocopy when so commanded).

The CRT display is a serial- or parallel-loaded device having its own memory for storing what is on the screen. Data entry is usually via a keyboard having switch contacts and logic circuits. To enter data onto the screen in a particular place, the operator moves a cursor, which is a blinking underline or other identifiable character, to the location he desires. There is usually a key for advancing the cursor one space, backspacing, advancing one line, and going up-screen one line. He can then type whatever characters he wishes. The computer also has control of the cursor and can respond in a similar manner. This typing, receiving a computer response, typing, etc., is called an interactive system since the operator interacts with the computer. It contrasts with a batch system where the programmer submits his cards to the computer operator and comes back later to receive his line-printer printout.

Eventually, the lines of data on the CRT will be at the bottom of the screen. When the next line is required, the CRT automatically "rolls up" one line, moving the entire display up one line (the top line is lost) and displaying the newest line at the bottom.

Magnetic Auxiliary Storage Techniques

All a computer's storage cannot consist of only core or IC memory; they are much too expensive. Therefore, most large systems have some method or methods of storing data on a surface coated with ferromagnetic material, usually magnetic tape, magnetic disk, or magnetic drum. These three differ in access time, the time it takes to obtain any particular character within the storage medium. The drum has the least access time, approximately 8 ms, the disk about 80 ms, and the magnetic tape can take on the order of minutes. However, the tape is the least expensive, the disk next, and the drum the most expensive of the three per bit of storage. Thus, access time is inversely related to cost.

Data is stored on these magnetic devices by saturating the magnetic surface in one direction and then saturating the surface in the opposite direction as the medium moves past the record head (Fig. 10-12). Similarly, data is read by moving the medium past a read head, which picks up the changes in magnetic flux on the surface and converts them into electrical signals.

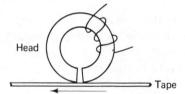

Head

Tape

FIG. 10-12 Magnetic Recording

Two types of recording methods can be used: return to zero (RZ) and non-return to zero (NRZ). In the return to zero method [Fig. 10-13(a)], the signal always returns to 0 V (or flux or amps) after a 1. A 0 is transmitted as 0 V. Note that the signal does indeed return to 0 after every bit.

The RZ system has the disadvantage of requiring some type of external timing to determine when a bit starts and when it finishes. Note that if a string of 0's is transmitted, there is great difficulty in selecting the third 0, for example, unless such a timing signal is present.

The NRZ method [Fig. 10-13(b)] returns to 0 only when a logical 0 is to be transmitted. This decreases the required frequency response of the system, allowing greater packing densities. Again, though, timing must be supplied externally. The NRZI method [Fig. 10-13(c)] requires that the flux change direction in the middle of the bit if a 1 is to be transmitted and no flux changes take place when a 0 is transmitted.

There are two popular methods that transmit the clock along with the data. In the phase modulation system [Fig. 10-13(d)] the flux always changes in the middle of a bit. If it changes from negative to positive, a 1 is being transmitted, and if positive to

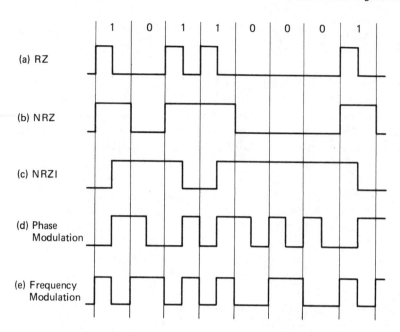

FIG. 10-13 Magnetic Recording Techniques

negative, a 0. By **OR**ing these changes in the middle of the bit, the timing can be extracted from the waveform. In the frequency modulation system, the signal always changes at the bit boundary. If a 1 to be transmitted, there is also a reversal in the middle of the bit. If a 0 is to be transmitted, there is no reversal. Note that the timing can be extracted from the bit-boundary changes.

Magnetic Tape

Magnetic tape is used because of its low cost and permanent storage qualities. A tape 2400 feet (ft) in length packed at 800 characters per inch (in.) can contain over 20 million characters. Further, purchasing of additional tapes can increase the storage capacity of a system indefiniely.

The tape is manufactured from $\frac{1}{2}$-in.-wide plastic ribbon, coated with a thin layer of magnetic material. Data are usually phase-encoded on nine parallel tracks (seven are also available) at packing densities of 200, 556, 800, and 1600 characters per inch. A "record" on tape consists of a series of characters followed by a check sum, followed by a $\frac{3}{8}$-in. length of black tape called an *interrecord gap*. A file is composed of records, followed by a 3.7-in. length of blank tape called an *end of file gap*, followed by a character called an *end of file mark*. These blank gaps in the tape allow the tape transport time to stop and start. Although incremental recorders are available that allow the stopping between characters, most can stop only between records.

The tape transport is a rather remarkable device. To reduce the amount of inertia the capstan has to control, the tape is looped into two vacuum columns

(Fig. 10-14). In this manner, the capstans and pinch rollers have only the inertia of the tape to overcome, enabling them to start and stop in less than 1 ms. Sensors monitor the length of the tape loop in the vacuum columns and control servomechanisms that move the supply reel and take-up reel motors.

Cassette magnetic tape recording is also used, providing storage at up to 3600 bits per inch and transfer rates 500 bits to 18 kilobits per second.

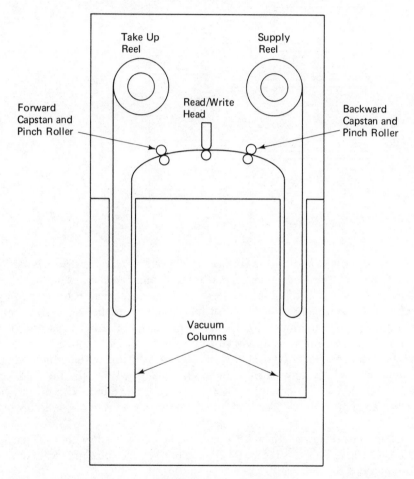

FIG. 10-14 Tape Transport

Magnetic Disk

The magnetic disk is used because of its fast access time (faster than tape) and lower cost (lower than drum). Two types of disk systems are available: removable disks, called disk packs, and fixed disks. A system with removable disks can have unlimited storage space.

Several disk surfaces are usually available on the same drive (Fig. 10-15), each with its own head. These heads float over the surface on a cushion of air, reading from and writing onto the disk. All heads are driven simultaneously by the head positioning mechanism toward or away from the center.

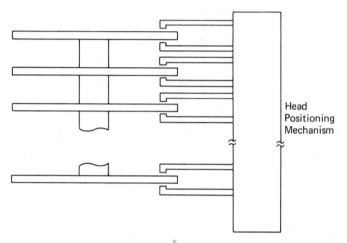

FIG. 10-15 Disk Storage

Information is stored on the surfaces in concentric circles called *tracks*. When the head system is positioned over a particular track, number 29 for example, the heads are addressing number 29 of all the tracks of all the surfaces. This is referred to by the programmers as a *cylinder*, cylinder 29 in this case. One particular disk system contains 10 disks (20 surfaces) with 203 tracks per surface, each track capable of storing over 7000 characters, for a maximum storage capacity of 29.5 million characters per disk pack. Of the 203 tracks on each surface, 200 are available to the programmer and 3 are spare tracks that may be substituted for a damaged one (by the field service engineer). The disk drive operates at 2400 revolutions per minute (r/min), providing an average access time of 87.5 ms and a maximum transfer rate of 156 kilobytes per second.

Disk packs are now available that contain their own heads within the pack, one head for each track. This cuts the access time to on the order of 20 ms, while retaining the portability of the disk pack.

A device called the floppy disk is also available. About $\frac{1}{16}$ in. thick, it has one or two magnetic recording surfaces on a flexible plastic substrate contained within a protective cartridge. Rotating at 300 r/min, it can provide more than 2 million bits of storage per surface with access times of 25 to 120 ms. Because of its low cost and relatively fast access time (compared with cassette), it is a very popular storage device in many systems for everything from scratch-pad storage to mass storage to data entry recording instead of a key punch.

Magnetic Drums

Magnetic drums are being used for fast-access high-capacity storage. The drum is a cylinder with a magnetic coating; it rotates at a constant speed, some to 14,000 r/min. The drums usually contain one head per track with 800 track systems available at a transfer rate of 800,000 bytes per second and typical access times of 8 ms.

Because of the drum's exacting mechanical tolerances and the plummeting cost of semiconductor memory, coupled with the flexibility of disk packs, the drum storage technique is less popular than it was a few years ago.

10-8 SUMMARY

All sizes and types of computer systems are available to the designer today. The computer system consists of the six units: input, output, control unit, memory, arithmetic unit, and accumulator. The control unit, arithmetic unit, and accumulator form the processor subsystem. A computer executes instructions by fetching them from a memory into the IR, decoding them, and executing the operation. Computer instructions fall into two categories: memory reference and nonmemory reference. The memory subsystem has a memory address register (MA), which contains the address to be examined, and a memory data register (MD), used for storing data to be written into memory or the data read from memory.

The MC6801 microcomputer has seven building blocks within one chip: the clock, processor, RAM, ROM, serial IO, parallel IO and timer.

A wide variety of input/output devices is available, including tape readers and punches, teletypewriters, card punches and readers, line printers, and CRT terminals. Mass storage is available on such devices as magnetic tape, disk, and drums.

10-9 PROBLEMS

10-1. What registers does the control unit of a computer contain?

10-2. Write a FIC-1 program to add two numbers together. If the result is 0, the accumulator must contain a 0 upon completion; if the result is not 0, it must contain a 1.

10-3. Write a FIC-1 program to go through a list of five numbers and find out how many times the number 0010 1001 appears.

10-4. Assume the FIC-1 had a subtract instruction. Using the block diagram in Fig. 10-3, describe how it could be executed. Use notation similar to that used in Table 10-4.

10-5. What is an MPU and what is its purpose?

10-6. A program can be written to provide a string of pulses at regular intervals. Why, then, is a timer included on the MC6801?

10-7. What are the relative advantages and disadvantages of serial versus parallel communication?

10-8. How many bytes of external memory may be added to the MC6801 when in the expanded, non-multiplexed mode?

10-9. The expanded, multiplexed modes use 16 address lines, giving 65k bytes total of memory space. Suggest a method by which this might be substantially increased.

10-10. Why does the MC6801 have two input pins for $+5$ V?

10-11. A computer can input from an external device to tell if it has a word ready to be received by the processor. Thus, a program can be written that periodically checks this status. If this is so, why are interrupts necessary?

10-12. What are the advantages and disadvantages of direct versus extended addressing modes?

10-13. Why would a LOAD INDEX REGISTER IMMEDIATE instruction be three bytes long whereas a LOAD A REGISTER IMMEDIATE instruction is only two bytes long?

10-14. What is the advantage of indexed mode instructions over extended mode instructions?

10-15. What is the advantage of using a floppy disk over that of using magnetic tape for mass storage?

10-16. What is the disadvantage of RZ recording?

10-17. Compare the access times of magnetic tape, magnetic drum, and magnetic disk mass storage.

ANALOG/DIGITAL CONVERSION

11

11-1 ANALOG MEETS DIGITAL

There are many system problems that require connecting a digital portion of the system to an analog component. This "meeting of the circuits" is called an *interface*. There are two problems associated with this interface that must be met (and conquered): (a) the conversion of digital signals to analog, and (b) the conversion of analog signals to digital. In this chapter we shall examine the methods used for these conversions.

Expressing Analog Voltages in Binary

Any analog voltage can be expressed as a binary word by assigning voltage weights to each bit position. Consider a four-bit word. Voltage values of 8, 4, 2, and 1 could be assigned to each bit position as follows:

Binary Word	Voltage
0000	0
0001	1
0010	2
0011	3
0100	4
. . .	. . .
1101	13
1110	14
1111	15

Note that each successive binary count represents 1/15th of the entire voltage. Thus, if a voltage of 13.6 V must be represented, 1110, a binary 14, is as close as we could come. A graph of these voltages versus their binary equivalent is shown in Fig. 11-1. Note that this is a step process—voltage is resolvable only into discrete binary words.

If this concept were extended to eight bits, each successive binary count would be equal to 1/255th of the total. We can say the resolution of this system is 1/255 or 0.392% of full scale. In general, the percent resolution is

$$\% \text{ Res} = \frac{1}{2^N - 1} \times 100$$

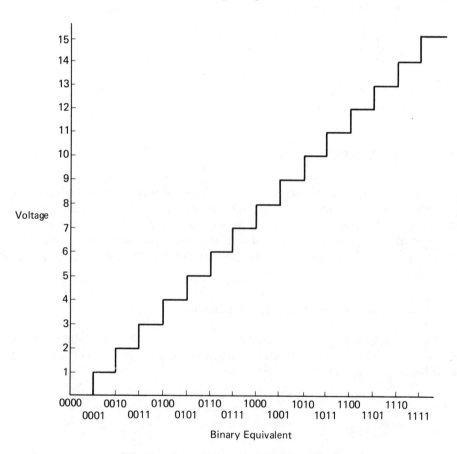

FIG. 11-1 Expressing Analog Voltages in Binary

Example 11-1: What is the percent resolution of a five-bit digital to analog (D/A) converter?

SOLUTION:

The maximum number that can be represented using five bits is $11111_2 = 31_{10}$. Therefore, the percent resolution is

$$\% \text{ Res} = \frac{1}{2^N - 1} \times 100 = \frac{1}{2^5 - 1} \times 100 = 3.23\%$$

Table 11-1 details the resolution of several digital to analog converters. Since this resolution problem introduces error into the system, it is desirable to use as many bits as possible. However, component value tolerances must be less than the percent resolution to introduce a minimum of error. Therefore, the greater the number of bits used, the more precise the resistors in the D/A must be. The trade-off is precision of output versus cost.

TABLE 11-1. D/A and A/D Converter Resolution

Number of Bits	Percent Resolution
4	6.67
6	1.75
8	0.392
10	0.976
12	0.0244
14	0.00610
16	0.00153

Example 11-2: A six-bit A/D converter has a maximum precision supply voltage of 20 V. Provide the following information for the unit:

1. What voltage change does each LSB represent?
2. What voltage does 100110 represent?

SOLUTION:

Each bit represents $1_2/111111_2$ or 1/63rd of the total of 20 V. Therefore, each LSB change represents

$$E = \frac{1}{63} \times 20 = 0.317 \text{ V}$$

The binary number 100110 equals 38_{10}. Therefore, a 100110 represents

$$E = \frac{100110_2}{111111_2} \times 20_{10} = \frac{38}{63} \times 20 = 12.06 \text{ V}$$

11-2 DIGITAL TO ANALOG CONVERSION

To convert a digital signal to analog, it is necessary to treat each bit in this weighted manner. A block diagram of such a device is shown in Fig. 11-2. The reference voltage source feeds a precisely regulated voltage to the voltage switches. Upon receipt of a CONVERT signal, binary data is clocked into the register, with each bit assigned a weighted value of current or voltage. These binary signals from the input register next feed voltage switches which provide one of two possible outputs: 0 V or the precision voltage source voltage. Thus, they are equivalent to an ordinary SPDT

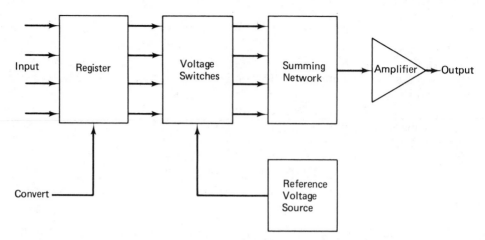

FIG. 11-2 Block Diagram, Digital to Analog Converter

switch controlled by the binary signal from the register. The switches feed a resistive summing network which converts each bit into its weighted current value and sums them for a total current. This total value is then fed the amplifier, which performs two functions, current to voltage conversion and scaling, so that the output voltage of the D/A converter will be the proper value.

Input Register

The input register is a parallel-in, parallel-out device. The CONVERT signal is used to clock the input data into the register where it is stored until the next CONVERT signal is received.

Voltage Switches

The simplest scheme for a voltage switch is an integrated circuit containing a field effect transistor (Fig. 11-3). With a high logic level on the input, Q_1 will cut off and Q_2 turn on, grounding the output. With a low level on the input, Q_2 will cut off and Q_1 turn on, connecting V_{REF} to the output. Thus, the device outputs V_{REF} or GND to the resistance network.

Resistive Summing Networks

The purpose of the resistive network is to weight and sum the binary currents. In the network shown in Fig. 11-4, resistor values are weighted inversely with their current values. Note that the most significant bit is bit A, since it supplies the most current. For simplicity, assume the register outputs plus 8 V and ground. If $ABCD$ were 0000, the output would be 0 mA. If it were 1010, the output would be $8 + 0 +$

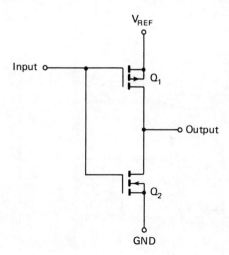

FIG. 11-3 Voltage Switch

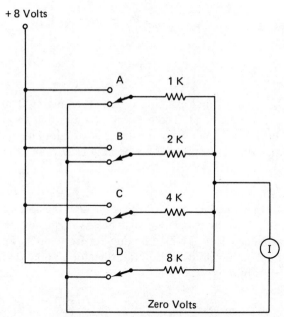

FIG. 11-4 Weighted Resistor Network

$2 + 0 = 10$ mA. If it were 0110, then the current would be $0 + 4 + 2 + 0 = 6$ mA. Note that the output current represents the decimal equivalent of the binary number expressed in milliamperes.

Although this weighted resistive network is very easy to understand, it has the disadvantage of requiring many different values of very low-tolerance resistors. The network shown in Fig. 11-5 is the one usually used; it is called a *ladder network*, probably because it looks like one. It can best be analyzed by superposition, determining what contribution to the total current each bit position supplies.

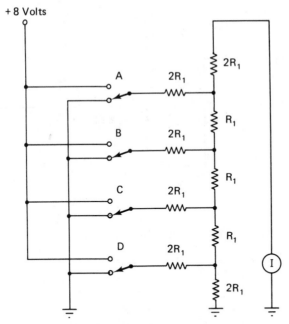

FIG. 11-5 Ladder Network

Let us first examine switch A, assuming it to be switched to $+8$ V, and B, C, and D to ground. The network becomes that shown in Fig. 11-6. Note that looking down from point E the resistance is $2R_1$; looking up from point E it is also $2R_1$. Thus, the total resistance seen by point A is $3R_1$, and the total current is

$$I = \frac{E}{R} = \frac{8}{3R_1}$$

Of this, one-half or $4/3R_1$ goes through the ammeter.

Now let us analyze switch B as it is connected to $+8$ V and the other switches grounded (Fig. 11-7). Looking up from point F, the total resistance seen is $2R_1$. Looking down, it is also $2R_1$; therefore, point B sees $3R_1$, and the total current supplied by B is

$$I = \frac{E}{R} = \frac{8}{3R_1}$$

I_1 is one-half of this value or $4/3R_1$, and the current through the ammeter one-half of this value, or $2/3R_1$. Therefore, switch A contributes $4/3R_1$ to the ammeter current and switch B, $2/3R_1$; note that switch B contributes one-half as much current as switch A. Similarly, switch C contributes one-fourth the current A does, and switch D, one-eighth. Therefore, the switches contribute weights of $1:2:4:8$, the same as is done by the weighted resistive network shown in Fig. 11-4.

There are other codes that can be used with a D/A converter: BCD and XS3 are examples. Two methods are available for converting BCD to the proper current values to feed the amplifier.

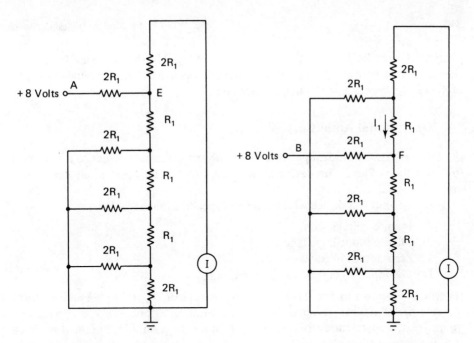

FIG. 11-6 Switch *A* Energized   **FIG. 11-7** Switch *B* Energized

The first method is used with relatively low-resolution D/A converters. In this method, a weighted resistor network is used [Fig. 11-8(a)], with the weight of the 10's resistor being one-tenth that of the unit's resistor. However, when the input

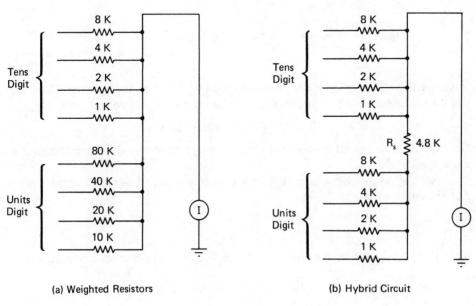

(a) Weighted Resistors   (b) Hybrid Circuit

FIG. 11-8 BCD Resistor Networks

consists of many BCD digits, the configuration of Fig. 11-8(b) can be used, minimizing the number of unique resistor values required. Each BCD digit would have an R_s of 4.8 kΩ between it and the next higher digit.

Operational Amplifiers

The purpose of the amplifier is to (a) convert the input current to a voltage and (b) scale the voltage. The device usually employed is an operational amplifier (op amp).

An op amp is a device whose assumed specifications include

A. Infinite voltage gain.
B. Infinite input impedance.
C. Zero output impedance.
D. The input voltage is virtually ground.

Its symbol is shown in Fig. 11-9, with a series resistor, R_s, of 10 kΩ and a feedback resistor, R_f, of 100 kΩ. Assume the device is an inverter and $+1$ V is applied to the input. Next, analyze the current at point A on the diagram. Since point A is virtually 0 V, the current through R_s is

$$R = \frac{E}{I} = \frac{1}{10} = 0.1 \text{ mA}$$

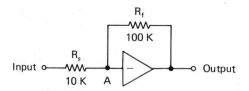

FIG. 11-9 Operational Amplifier

But there is no current flow into or out of the $(-)$ input on the op amp. Therefore, this 0.1 mA must be coming from R_f. But if R_f is 100 kΩ, its voltage drop is

$$E = IR = 0.1 \times 100 = 10 \text{ V}$$

Since point A is a virtual ground, the output voltage must be -10 V and the gain of the amplifier 10.

Next, let us develop a general formula for this critter. Assume an input voltage E_{in}. The current through R_s is

$$I = \frac{E_{in}}{R_s}$$

Since this is the same as the current through R_f, the voltage drop across R_f is

$$E_0 = IR = \left(-\frac{E_{in}}{R_s}\right)(R_f)$$

The minus sign indicates the current through R_f is in the opposite direction from the current through R_s.

Next, let's develop a formula for voltage gain. Since

$$E_o = -\frac{E_{in}R_f}{R_s}$$

voltage gain, A_v, is

$$A_v = \frac{E_o}{E_{in}} = -\frac{R_f}{R_s}$$

Therefore, the voltage gain of an operational amplifier is dependent only on its external resistors.

Example 11-3: Compute the gain for the op amp shown in Fig. 11-10. What is its output voltage if E_{in} is 3.2 V?

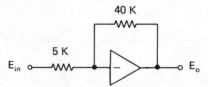

FIG. 11-10 Example 11-3

SOLUTION:

Since $A_v = R_f/R_s$,

$$A_v = \frac{R_f}{R_s} = \frac{40}{5} = 8$$

The voltage gain is, therefore, 8. Since the input voltage is 3.2 V,

$$A_v = \frac{E_o}{E_{in}}, \qquad E_o = AE_{in} = 8 \times 3.2 = 25.6 \text{ V}$$

The op amp can also be used as a current summer, a characteristic mighty attractive for a D/A converter. Since the current into point A of Fig. 11-11 must equal the current leaving,

$$-(I_1 + I_2 + I_3) = I_f$$

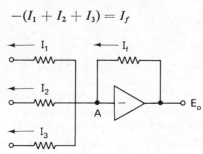

FIG. 11-11 Current Summing

But

$$E_o = I_f R_f = -(I_1 + I_2 + I_3)R_f$$

Note that this is an algebraic sum of currents.

Example 11-4: Compute the output voltage for the circuit in Fig. 11-12.

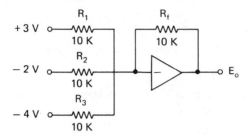

FIG. 11-12 Example 11-4

SOLUTION:

The currents through the series resistors are

$$I_1 = \frac{E_1}{R_1} = \frac{3}{10} = 0.3\,\text{mA}$$

$$I_2 = \frac{E_2}{R_2} = -\frac{2}{10} = -0.2\,\text{mA}$$

$$I_3 = \frac{E_3}{R_3} = -\frac{4}{10} = -0.4\,\text{mA}$$

The output voltage is

$$\begin{aligned}
E_o &= -(I_1 + I_2 + I_3)R_f \\
&= -(0.3 - 0.2 - 0.4)(10) \\
&= 3.00\,\text{V}
\end{aligned}$$

Note that this circuit is also a voltage summer, since the voltage gain for each branch is

$$A_v = -\frac{R_f}{R_s} = -\frac{10}{10} = -1$$

Therefore, the sum of the voltage inputs is

$$\text{Sum} = +3 - 2 - 4 = -3.00\,\text{V}$$

and with an A_v of -1, the output voltage is $-(-3)$, or $+3$ V.

The op amp used for the D/A converter is a combination current to voltage converter and voltage amplifier. Assume switches $ABCD$ of Fig. 11-13 are set to 1011. (This circuit is used for its simplicity of analysis. In practice, the ladder network is usually used.) The current leaving junction E would be

Switch A:	$I = E/R = 8/1 =$	8.00 mA
Switch B:	$I = E/R = 0/2 =$	0.00 mA
Switch C:	$I = E/R = 8/4 =$	2.00 mA
Switch D:	$I = E/R = 8/8 =$	1.00 mA
Total current:		11.00 mA

With a total current of 11 mA out of the junction, the output voltage is

$$E_o = I_f R_f = -11 \times 1 = -11.0 \text{ V}$$

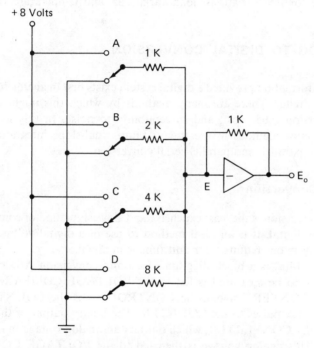

FIG. 11-13 Summing Network and Amplifier

Example 11-5: Assume switches $ABCD$ of Fig. 11-13 are 0110. Compute output voltage.

SOLUTION:

The currents out of junction E are

$I_A = E/R = 0/1 =$	0.00 mA
$I_B = E/R = 8/2 =$	4.00 mA
$I_C = E/R = 8/4 =$	2.00 mA
$I_D = E/R = 0/8 =$	0.00 mA
$I_{tot} =$	6.00 mA

Therefore,

$$E_o = I_f R_f = -6 \times 1 = -6.00 \text{ V}$$

Reference Voltage

The reference voltage source must have less percent error than the percent resolution dictated by the number of bits used. For example, a 12-bit D/A converter must have a reference voltage of better than 0.0244% regulation. This requires design to compensate for line variations, load variations, and temperature variations.

11-3 ANALOG TO DIGITAL CONVERSION

Much of the information provided a digital system exists first in analog form, requiring conversion to digital. There are many methods by which this might be done, with varying conversion rates, cost, and susceptibility to noise. In this section we shall examine the major methods in use today; ramp, dual-slope integration, successive approximation, parallel, and parallel/serial conversion.

Ramp Conversion

Ramp conversion is the least expensive and slowest method of converting analog information to digital. It is an ideal method to use in a digital voltmeter where the number of conversions required per unit time is minimal.

Figure 11-14(a) is a block diagram of a ramp conversion A/D converter. The INPUT voltage to be measured is fed to the VOLTAGE COMPARATOR: Upon receipt of the CONVERT signal, the CONTROL resets the COUNTER to 0 and then supplies clock pulses to the COUNTER. The binary output of the COUNTER is fed to the D/A CONVERTER, which outputs an analog voltage in response to its digital input. This analog voltage is then fed to the VOLTAGE COMPARATOR which compares the output of the D/A CONVERTER with the analog input. As soon as the D/A input exceeds the INPUT voltage, the COMPARATOR signals the CONTROL circuit, and it stops the COUNTER. The binary number in the COUNTER then represents the voltage of the input signal. The CONTROL circuit will also output a POLARITY signal, indicating whether positive or negative, and an OVERFLOW signal if the input signal exceeds the highest possible voltage of the D/A CONVERTER.

Figure 11-14(b) illustrates a waveform converted using this method. As can be seen, most of the time between samples is taken up by the counter. Because of this, only three samples can be taken over the entire waveform. If these samples were then converted back to analog, the resultant waveform would be as shown by the dashed lines.

A modification of the converter results in much improved operation. By making

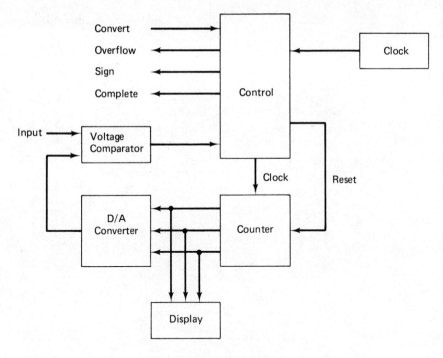

(a) Block Diagram

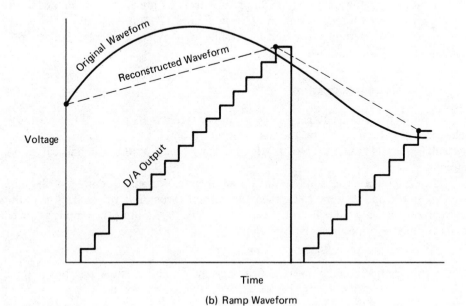

(b) Ramp Waveform

FIG. 11-14 Ramp A/D Conversion

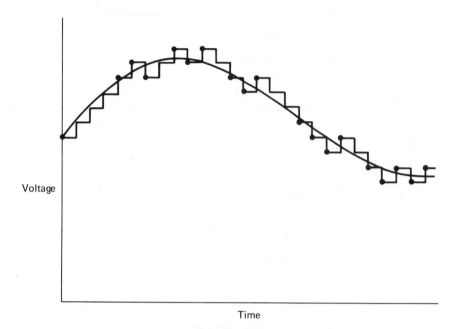

Voltage

Time

(c) Up/Down Ramp Waveform

FIG. 11-14 (continued)

the counter an up/down counter, it can proceed from its previous analog point to the next analog point without having to be reset. Figure 11-14(c) illustrates a waveform converted by this method. Note that this results in many more samples, better representing the original waveform.

Dual-slope Integration

One of the disadvantages of the ramp converter is its susceptibility to noise. This problem is overcome by the dual-slope method (Fig. 11-15). The core of this system is the INTEGRATING AMP, an operational amplifier connected as an integrator.

Let us take a closer look at this op amp configuration. If point *A* of the amp were at +1 V and *R* were 1 kΩ, then the current through *R* would be 1 mA. But this current can only come from the capacitor, since the input of the amp is assumed to have infinite impedance. However, the formula for charge on a capacitor is

$$Q = CV$$

where Q is charge in coulombs (C), C is capacitance, and V is voltage. But observe that current is defined as charge per unit time (coulombs per second). If 1 A is flowing through a circuit, this represents 1 C/s. Now let's plug this concept into the above formula. If Q is increasing at 1 C/s (thus, 1 A is flowing), then the voltage must be

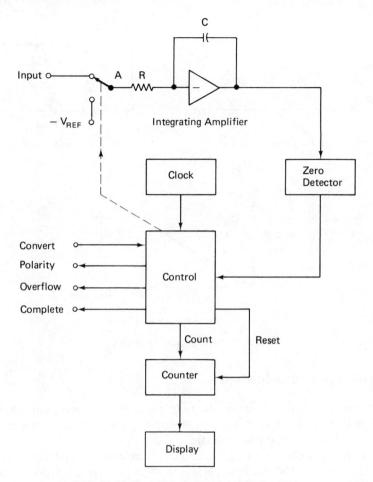

FIG. 11-15 Dual Slope Integration A/D Converter

increasing at some linear rate. Therefore, constant current through a capacitor will result in a ramp voltage across that capacitor (Fig. 11-16).

From the foregoing discussion, we can conclude that when the INTEGRATING AMP is connected to a voltage source, it will produce a ramp voltage at its output. Upon receipt of a CONVERT signal, the AMP is connected to the INPUT for a predetermined length of time. The COUNTER is then turned on, and the AMP is switched to the −REF voltage, resulting in current flow in the opposite direction through R, discharging C. This produces a negative ramp at the AMP output, ultimately crossing through 0 V. When 0 V crossover is detected, the COUNTER is stopped and its binary value represents the INPUT voltage.

The dual-slope integration method has the advantages of low cost, relatively short conversion time (approximately 50 ms), and good noise immunity, since the input is really an average of the waveform while it is connected to the op amp.

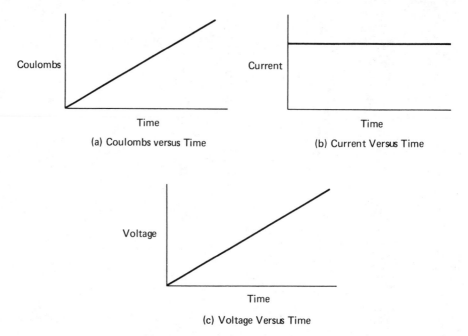

(a) Coulombs versus Time   (b) Current Versus Time

(c) Voltage Versus Time

FIG. 11-16 Integrator Waveforms

Successive Approximation

With the advent of computers, shorter conversion times were required—on the order of microseconds, rather than milliseconds. The successive approximation method was devised to meet this challenge.

Figure 11-17(a) is a block diagram of the system, and Fig. 11-17(b) illustrates the output of the D/A CONVERTER compared with the INPUT voltage. Assume a five-bit conversion must be made for a 26.2-V signal using a maximum reference voltage of 64 V. Referring to the flow diagram, the REGISTER is first set to 0, then a 1 is placed in its MSB. This is converted by the D/A and fed to the COMPARATOR where it is compared with the INPUT voltage. If the D/A output exceeds the INPUT, the bit is set to a 0; if the D/A is less than the INPUT, the 1 is retained in the REGIS-TER. In this case, the output of the D/A is 32 V, resulting in this bit being set to 0.

Next, a 1 is placed in the next bit to the right, resulting in an output of 16 V from the D/A. Since this is less than the INPUT, the 1 is retained in this bit position. Placing a 1 in the third position results in a 24-V output, again less than the INPUT. Therefore, the 1 is retained. Placing a 1 in the fourth position results in an output of 28 V, exceeding the INPUT. Therefore, bit 4 is set to 0. A 1 in bit 5 results in 26 V, less than the INPUT; it is therefore retained. A 1 in bit 6 results in 27-V output, exceeding the INPUT; it is therefore set to 0. Thus, the resultant binary word is 011010, representing 26 V.

The successive approximation method is very fast; many units convert in less than 250 ns/bit. However, it is more expensive than the ramp methods.

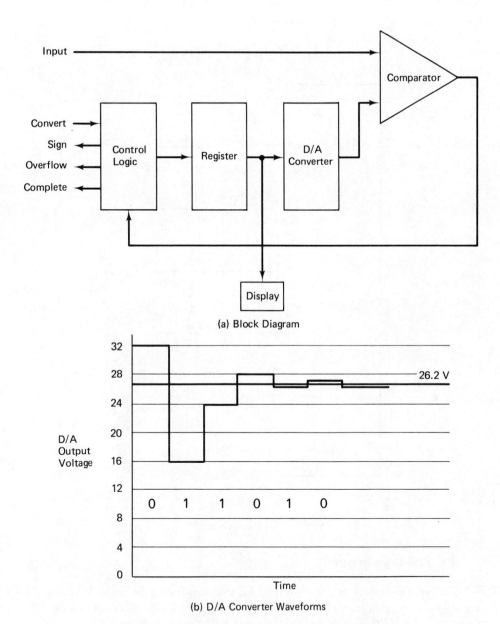

(a) Block Diagram

(b) D/A Converter Waveforms

FIG. 11-17 Successive Approximation A/D Converter

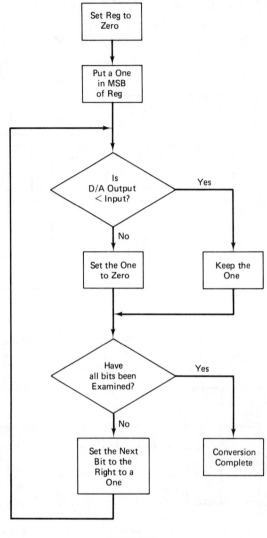

Parallel Conversion

Parallel conversion is the easiest of all the conversion methods. It is also the fastest and most expensive. Figure 11-18 illustrates this method for a three-bit A/D converter. The INPUT is fed to all seven voltage level COMPARATORS. If the INPUT were 3.23 V, the outputs of COMPARATORS *A*, *B*, *C*, and *D* would all be 0's, and those from *E*, *F*, and *G* would be 1's. The ENCODER would then translate this into a three-bit code and clock it into the REGISTER.

The big disadvantage of this system is cost, for an *N*-bit converter requires

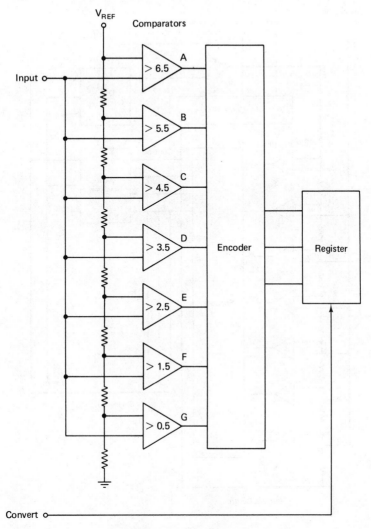

FIG. 11-18 Parallel Conversion

$2^N - 1$ comparators. Therefore, a 10-bit converter would require $2^{10} - 1$ or 1023 comparators. However, in applications requiring little resolution and great speed, it is very useful.

Parallel/Serial Conversion

The speed of parallel conversion and the low cost of successive approximation (also called serial conversion) can be combined by using the parallel/serial method (Fig. 11-19). In this system, the CONVERT pulse causes the CONTROL section to

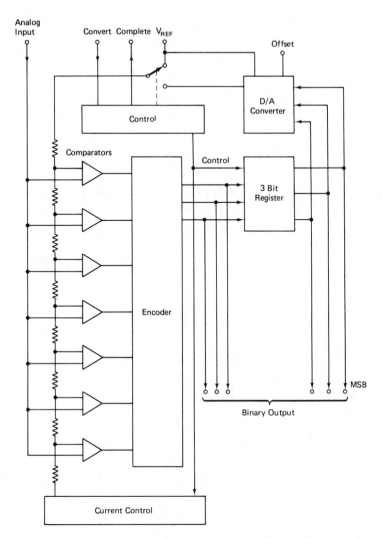

FIG. 11-19 Parallel/Serial A/D Converter

apply V_{REF} to the COMPARATOR divider string and the CURRENT CONTROL such that they represent the voltages expressed by the three most significant bits of the binary output word. For example, assume six bits, with a V_{REF} of 64 V. The inputs would represent 8, 16, 24, 32, 40, 48, and 56 V, reading from bottom to top. The COMPARATORS and ENCODER then output the three most significant bits of the word, and this is clocked into the REGISTER. This is then fed to the D/A CONVERTER. At this time, the CONTROL switches the COMPARATOR resistors to the D/A OUTPUT, while reducing string current via the CURRENT CONTROL.

The three least significant bits can now be parallel-converted and this result fed to the output via the ENCODER.

As an example, assume a 26-V input must be converted using our V_{REF} of 64 V. The first step would determine the three most significant bits to be 011, since it is greater than 24 and less than 32 V. This 011 is fed to the D/A, which has a $+8$-V offset designed into it, causing it to output 32 V (8 above the 24). The CONTROL then switches the COMPARATOR divider string to the D/A output and restricts the current in the divider string so that the voltages, from top to bottom, are 31, 30, 29, 28, 27, 26, and 25. The least significant bits can now be formed from the outputs of the COMPARATOR and ENCODER; in this example the three LSBs are 010. The entire output word is, therefore, 011010. This can be clocked into an OUTPUT REGISTER (not shown) if desired.

This same design can be extended to any number of bits by providing multiple registers and switching circuits. In the example shown, a two-step procedure provided six bits of accuracy; using a straight serial approach, it would require 6 steps, and using counting, up to 63 steps.

11-4 CONVERTER SPECIFICATIONS

Both A/D and D/A converters are available in a wide range of prices and specifications. To intelligently select the proper device for the job, these specifications must be carefully analyzed. The D/A specifications will be analyzed first and then the A/D specs.

D/A Resolution

The resolution of a D/A converter is defined as the smallest observable change in the analog output that can be effected by a single step change in the digital input. The percent resolution is this change expressed as a percent of full-scale output.

Example 11-6: What is the resolution and percent resolution of a 12-bit D/A converter whose output varies between -50 and $+50$ V?

SOLUTION:

Since a 12-bit converter can provide $2^{12} - 1$ or 4095 non-0 states, the resolution is

$$\text{Res} = \frac{E}{2^N - 1} = \frac{50 - (-50)}{2^{12} - 1} = \frac{100}{4095} = 0.0244 \text{ V}$$

This means that the smallest analog change in the output (a change in one LSB) is 0.0244 V.

The percent resolution is

$$\% \, \text{Res} = \frac{1}{2^N - 1} \times 100 = \frac{100}{4095} = 0.0244\%$$

This means that a change of one LSB in the input will result in a change of 0.0244% of the full-scale output voltage.

D/A Accuracy

There are two specifications for accuracy: relative accuracy, also called *linearity*, and differential linearity. Relative accuracy is the maximum deviation of the output of the D/A converter compared with the ideal line (Fig. 11-20). Ideally, it should be $\pm\frac{1}{2}$ LSB. Differential linearity is the deviation of the output from an ideal one LSB change compared with the actual LSB change. If one bit is supposed to change the output voltage by 50 mV and it actually changes it by 51 mV, the differential linearity, expressed in percent, would be

$$\% \, \text{DL} = \frac{51 - 50}{50} \times 100 = 2.0\%$$

Ideally, this should be 0.

Monotonicity is defined as the quality of a D/A converter having no differential linearity problem; monotonicity thus implies $\pm\frac{1}{2}$ LSB accuracy.

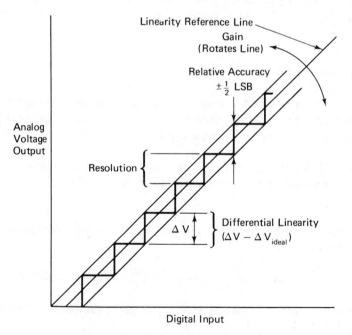

FIG. 11-20 D/A Converter Specifications

D/A Gain

Gain of a D/A converter is the voltage output divided by the equivalent voltage of the digital input at the linearity reference line. It can usually be zeroed out and is, therefore, not a serious problem.

D/A Speed

D/A speed is the amount of time necessary to settle to a desired accuracy. It is usually expressed as two specifications:

A. The amount of time necessary to change from one end of the range of digital codes to the other end.
B. The amount of time to change from one input code to the next sequential input code (a change of one bit).

A/D Resolution

A/D resolution is defined as the voltage input change necessary for a one-bit change in the output (Fig. 11-21). It can also be expressed as a percent.

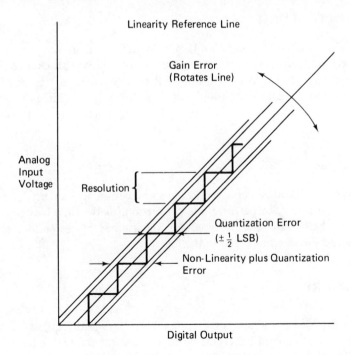

FIG. 11-21 A/D Converter Specifications

Example 11-7: A 10-bit A/D converter has an input voltage of -10 to $+10$ V. What is its resolution and percent resolution?

SOLUTION:

Its resolution is

$$\text{Res} = \frac{E}{2^N - 1} = \frac{10 - (-10)}{2^{10} - 1} = 19.5 \, \text{mV}$$

Its percent resolution is

$$\% \, \text{Res} = \frac{1}{2^N - 1} \times 100 = \frac{100}{2^{10} - 1} = 0.0978 \%$$

A/D Accuracy

The total accuracy of an analog to digital converter is limited by the $\pm \frac{1}{2}$ LSB quantization error and the other errors of the system. It is, therefore, defined as the maximum deviation of the digital output from the ideal linearity reference line (Fig. 11-21). Ideally, it would approach the $\pm \frac{1}{2}$ LSB.

Differential linearity is defined as the maximum amount of voltage change necessary to cause the digital output to change one bit, minus the ideal voltage change necessary to change one bit. Compare this with D/A accuracy. They are quite similar.

A/D Gain and Drift

A/D gain is defined the same as D/A gain and is usually not a problem, for it can be zeroed out. Drift, however, is the quality of a circuit to change parameters with time. Drift errors of up to $\pm \frac{1}{2}$ LSB will cause a maximum error of one LSB between the first and the last transition. Very low drift is quite difficult to achieve and increases cost of the device, with 5 ppm/°C being about the best commercially available.

A/D Speed

A/D speed is another very important specification. It can be defined in two ways: the time necessary to perform one conversion, or the time between successive conversions at the highest rate possible. It is determined by the settling time of the components and the internal speed of the logic.

11-5 SUMMARY

Much of the data fed a digital system is in analog form and requires conversion to digital. Conversely, many outputs must be analog in form and will require conversion from digital to analog.

D/A converters consist of an input register, voltage switches, current summing network, amplifier, and reference voltage source. Two types of resistive summing networks are used: weighted resistor values and the ladder configuration. They can be designed to convert binary, BCD, or any other code to analog.

The ramp A/D converter is the slowest and least expensive; the dual-slope integrator is relatively immune to noise. The successive approximation device converts in microseconds, whereas the parallel A/D converter can convert in nanoseconds. A hybrid, the parallel/serial A/D converter, takes advantage of the speed of the parallel device and low cost of the serial converter.

Both A/D and D/A converters have specifications of resolution, accuracy, speed, and gain; the A/D also has the problem of drift.

11-6 PROBLEMS

11-1. Each step of an eight-bit D/A converter represents 0.2 V. What does a code of 01001101 represent? How about 01100110?

11-2. Each step of a 10-bit D/A converter represents 0.025 V. What does a 0010010011 represent? How about 0101010101?

11-3. What is the percent resolution of a seven-bit A/D converter?

11-4. What is the percent resolution of a 13-bit A/D converter?

11-5. A system is required that represents 0–50 V in 10-bit code. What is the voltage increment that is represented by the LSB of this code?

11-6. A three-digit BCD code represents what percent resolution?

11-7. How many bits are required to achieve a resolution of 1 mV if full scale is 10 V?

11-8. What is the conversion time of a 10-bit successive approximation A/D converter if its input clock is 5 megahertz (MHz)?

11-9. Design a weighted resistor network for one digit of a 2421 BCD code.

11-10. How can a D/A converter be used for XS3 BCD code?

GLOSSARY

abacus: a counting device consisting of beads arranged in decimal weights

AC: accumulator

accumulator: a single word register that accumulates the partial results of arithmetic or logical operations

A/D: analog to digital

adder: a device that adds two binary numbers, giving a sum and a carry

address: a number representing the location of data within a memory

adjacency: a square of a Karnaugh map that may be combined with another square

algorithm: a set procedure for accomplishing an objective that always works

alphanumeric: alphabetic and numerical information

ALU: arithmetic and logic unit

American Standard Code for Information Interchange: a seven-bit binary code used to express alphanumeric information

analog: information expressed in measurable quantities, such as resistance, current, or voltage

analog to digital converter: a device that converts an analog input to a binary equivalent number

AND connective: a condition requiring all inputs to be true in order for the output to be true

arithmetic and logic unit: a logic device capable of performing both arithmetic operations (add, subtract, etc.) and logical operations (**AND, OR,** etc.)

ASCII: American Standard Code for Information Interchange

assertion level: the level (high or low) at which an event will occur

asynchronous circuit: a sequential circuit in which the next event occurs as soon as the present event is completed, without waiting for a clock pulse

AU: arithmetic unit; an ALU incapable of performing the logic functions

base: the number of single digits in a number system

BCD: binary-coded decimal

binary: a number system consisting of only the digits 0 and 1

binary-coded decimal: any of several methods of representing decimal digits as binary numbers

biquinary: a BCD code consisting of seven bits, two of which are 1's

bit: a binary digit

Boolean algebra: an algebra consisting of only three operations: **AND, OR,** and **NOT**

bus: parallel wires in which each wire represents one bit of a binary word

bush: the quality of a counter in which all states lead to the desired state

byte: a submultiple of a binary word, usually eight bits

category (of tabular reduction): the number of 1's in a binary number

cell: the smallest element of a memory capable of storing a 1 or a 0

character: a binary word used to express a number, letter, or punctuation

check sum: an error-detecting word in a sequence of binary data which contains the sum of all the data words

circle: symbol of an inverter

clear: to cause a flip-flop to go to the 0 state

clock: a pulse generator within a digital system to which all operations are synchronized

CMOS: complementary metal oxide semiconductor, an insulated gate field effect digital logic unit using both P and N MOS devices

code: a system of binary words used to represent decimal or alphanumeric information

code converter: a device that converts binary inputs from one code system into binary outputs in another code system

coincident current memory: a memory requiring additive currents in order to activate the memory cell

combinational logic: a logic system in which the output occurs in direct, immediate response to the input

comparator: a device that compares two inputs and provides an output indicating their relative magnitudes

complement: a word in which all 1's have been changed to 0's and all 0's to 1's

complementary ring counter: a ring counter in which the inverted output is fed into the input; also known as a Johnson counter

control system: a system in which the outputs are compared with a standard signal and an error signal fed to the input to modify the process

core memory: a memory using small toroids of ferromagnetic material to store binary data

cover: the property of a term to be a 1 upon input of the covered minterm

cycle counter: a device that outputs a specified counting sequence every time a control pulse is received

cylinder: the storage space formed in a multiple-surface disk system by addressing the same track on all surfaces

D/A: digital to analog

data selector: a device that selects one of several inputs depending on the control leads, and gates this input to the output; a multiplexer

decade: a quantity of 10

decade counting unit: a device that will output 1 pulse for every 10 received

decimal number system: a number system composed of 10 digits, 0 through 9

decoder: a device that will energize a particular output line or lines depending on the binary input

De Morgan's theorems: The Boolean formulas $\bar{A} \cdot \bar{B} = \overline{A + B}$ and $\overline{A \cdot B} = \bar{A} + \bar{B}$

demultiplexer: a logic circuit that gates its input to one of several outputs, depending on the binary word on its control leads

D flip-flop: a flip-flop whose output will be set to the value of the D input upon receipt of the next clock pulse

DG flip-flop: a flip-flop whose output may be changed to that on the D input lead by activating the G control lead

dibble-dobble: the double-up and add method of converting a binary number to decimal

differential linearity: the variation of the input voltage size of an **A/D** converter necessary to produce a change of one bit on the output

diffusion capacitance: the capacitance of a forward-biased semiconductor diode junction caused by unequal doping

digit: a single symbol of a number system

digital: a method of representing a number by discrete units

digital to analog converter: a device that converts a binary input to an analog output

digitize: to convert a signal to digital

diode-transistor logic: a family of logic using diodes and transistors as logic elements within the integrated circuit

DIP: dual-in-line package

direct logic: the method of sequential design in which the output is taken directly off the state flip-flops

don't care: a minterm for which the combinational logic may output either a 1 or a 0

double precision: a method of expressing a fixed-point binary number using two words instead of one

down counter: a counter that decreases its binary value by 1 every time a clock pulse is received

DTL: diode-transistor logic

dual-slope integration: a method of analog to digital conversion whereby an integrator is charged when connected to the analog input and then timed as it is discharged through the reference voltage

dynamic memory: a memory that will lose its data unless periodically refreshed

EBCDIC: Extended Binary Coded Decimal Interchange Code

ECL: emitter-coupled logic

edge-triggered flip-flop: a flip-flop whose output changes are dependent on only one edge of the clock pulse

8421 code: a binary-coded decimal number with bit weights of $8:4:2:1$

emitter-coupled logic: a high-speed nonsaturating logic family

error-correcting code: a code that detects and corrects its own errors

error-detecting code: a code that will detect an error in the data

excess-3 code: a binary-coded decimal code in which each word is equal to its binary value plus 0011

exclusive OR gate: a two-input gate requiring either input to be true (but not both) to obtain a true output

execute: (1) the process of performing the task specified by an instruction; (2) the portion of an instruction cycle in which the task specified by the instruction is performed

expander gate: a gate designed to increase the fan-in of a second gate

Extended Binary Coded Decimal Interchange Code: a method of expressing alphanumeric characters in eight-bit words based upon binary-coded decimal format

fan-in: the number of inputs to a particular gate

fan-out: the number of inputs a gate output is capable of driving

fetch: the portion of an instruction cycle during which the instruction is read from memory

flat pack: the integrated circuit package with flat leads

flip-flop: a logic device capable of storing one bit of data formed by cross-connecting two active devices

floating point: a method of expressing a number using a mantissa and the exponent of a base

frequency division multiplexing: a method of transmitting several channels over the same facility by assigning different frequency bands to each channel

frequency-modulated digital recording: a method of recording digital data in which a 1 is transmitted as two transitions and a 0 as one transition per bit position

full adder: a device capable of adding two binary numbers to a carry in and providing a sum and a carry out

gate: a simple logic element whose binary output is a Boolean **AND, OR,** or **XOR** of the input

Gray code: a binary code in which each successive number differs in only one bit position

half-adder: a logic device that adds two binary bits and provides a sum and carry out

Hamming code: a particular type of error-correcting code invented by Hamming

hexadecimal number system: a number system consisting of 16 unique symbols

Hollerith code: a binary code invented by Hollerith used for punched cards in expressing alphanumeric data

hybrid function: a sum-of-products or product-of-sums expression that has been reduced by factoring; three or greater combinational levels of logic

hysteresis: the quality of a logic device to recognize a 1 when it is in the 0 state at a higher voltage than that necessary for recognizing a 0 when it is in the 1 state

implicant: a reduced Boolean term in tabular reduction

implication table: a table used for reducing the number of asynchronous states to a minimum

incompletely specified function: a logic equation in which one or more minterms are don't cares

indirect logic: sequential logic in which the output is taken from gates, rather than the state flip-flops

instruction: a binary word that defines to the processor a task to be performed

instruction register: the register that holds the instruction currently being executed by the processor

integrator: an operational amplifier that outputs a ramp voltage when the input receives a dc value

interrecord gap: a $\frac{3}{8}$ in. length of blank tape between records that is used as a starting and stopping point for the transport

interrupt: a signal that causes the computer to cease what it is doing and go to another program

inverter: a logic device that outputs a 1 with a 0 input and outputs a 0 with a 1 input

IO: input-output

IR: instruction register

IRG: interrecord gap

JK flip-flop: a flip-flop capable of operating as a type T or type D, depending on the J and K leads

Johnson counter: a complementing ring counter

Karnaugh map: a system of reducing Boolean expressions using a matrix of 1's and 0's invented by Karnaugh

ladder network: a series/parallel resistor network used for D/A conversion

latch: a DG or RS flip-flop

least significant bit: the bit of a logic word that represents the least weighted value

level: the number of bit positions in a binary word

linearity: the digital output of an A/D converter compared with what the output should be

line driver: a logic element used to transfer data to a bus

line receiver: a logic element with input hysteresis designed to reject noise

literal: a letter in a position of a number

logic: a system of representing the validity of an output by analyzing inputs

logic diagram: a schematic diagram using logic symbols

LSB: least significant bit

LSI: large-scale integration

MA: memory address register

mapping: a system of reducing Boolean expressions using a matrix of minterms

master/slave flip-flop: a flip-flop composed of two internal FF's, one to receive the inputs (the master) and one to drive the outputs (the slave)

maxterm: an **OR** of single-lettered terms that includes all the variables

MD: the memory data register, also called the memory buffer

memory: a device for storing and retrieving binary words at locations called addresses

memory address register: the register that contains the address of the memory location currently being read or written into

memory data register: the register that contains the data to be written into the memory or the data that were just read from the memory

memory reference instruction: an instruction that requires an operand from memory or one that places data into memory

microprocessor: an integrated circuit capable of interpreting and, with the help of a RAM and input/output devices, executing instructions

Miller effect: the gain degeneration of an amplifier due to collector-base capacitance

minimization: the process of reducing a Boolean expression so that a minimum number of gates will be required for implementation

minterm: a product of single-lettered terms that contains all the variables

minterm designator: the decimal or binary number formed by assigning 0's to the complemented literals of a minterm and 1's to uncomplemented literals

modulo-n arithmetic: arithmetic in which all carries beyond $n - 1$ are discarded

modulus: the maximum possible number in an arithmetic system or register

monostable multivibrator: a device that will output a pulse of preset length every time it receives an input

monotonicity: the property of a D/A converter to maintain consistent analog voltage increments between binary increments

MOS: metal oxide semiconductor

most significant bit: the bit of a binary word having the most weight

MPU: microprocessor unit

MSB: most significant bit

MSI: medium-scale integration

multiple-output minimization: the process of reducing multiple equations to take advantage of terms that could be shared between the equations

multiplexer, binary: a multiple-input device that receives data on the input lead determined by the control leads and transmits them to the output

multiplexing: the process of transmitting multiple data channels over a single facility

NAND gate: an **AND** gate to which has been added an inverted output

natural BCD: a binary-coded decimal system using weighted values of $8:4:2:1$

negative logic: a logic system in which low levels are considered 1's and high levels 0's

nibble: a portion of a byte, usually four bits

noise immunity: the amount of noise a system can receive without being amplified beyond unity gain

noise margin: the voltage difference between the minimum voltage input guaranteed to be a logical 1 and minimum voltage output guaranteed to be a logical 1; also, the voltage difference between the maximum voltage input guaranteed to be a logical 0 and the maximum logical 0 voltage output

NOR gate: an **OR** gate to which has been added an inverter at its output

NOT connective: the property of negating a true function

NRZ: nonreturn to zero

NRZI: nonreturn to zero inverted

octal number system: a number system having eight digits, 0 through 7

one's complement: a system of binary arithmetic in which a negative number is expressed as the complement of its positive equivalent

one-shot: a monostable multivibrator

operand: one of the numbers upon which the computer operates

operational amplifier: an amplifier with an assumed infinite gain, infinite input impedance, and zero output impedance

OR connective: a condition in which any true input results in a true output

overflow: the exceeding of the maximum number a register can contain

parallel counter: a synchronous counting circuit

parallel D/A conversion: a D/A converter in which all bits are converted simultaneously according to their weights

parallel data: data in which each bit is represented by a single line; thus, 16 bits require 16 lines

parallel to serial converter: a device that converts parallel binary data to serial data

parity bit: a bit that is set to a 1 or 0 to make the total number of 1's in a binary word odd or even

PC: program counter

peripheral: input/output equipment or auxiliary storage equipment of a computer system

Petrick's method: a method of covering minterms by applying Boolean algebra to select prime implicants in tabular reduction

phase-modulated digital recording: a system of magnetic recording in which 1's are represented as positive transitions in the middle of a bit and 0's are represented as negative transitions

point, binary: the equivalent of a decimal point in binary

POS: product of sums

positional weighting: a number system in which the value of a digit is determined by its position with respect to the point

positive logic: a logic system in which positive voltages represent 1's and negative values 0's

preset: to cause a flip-flop to go to the 1 state

prime implicant: a tabular-reduced term incapable of being further reduced

processor: the portion of a computer that performs the arithmetic and provides system control

product of sums: a Boolean expression formed by **AND**ing **OR**ed term

program counter: a register that contains the next address from which an instruction will be taken by the computer

programmable read only memory: a read only memory with contents capable of being permanently changed by the designer

PROM: programmable read only memory

propagation time: the time for a level change on the input to result in a response at the output

pulse train: a repetitive series of pulses on a line

pulse-triggered flip-flop: a flip-flop that enters data into the master flip-flop at the leading edge of the clock pulse and transfers that data to the slave on the trailing edge of the pulse

punched cards: a card used to enter data by means of punched holes

quantization error: the voltage deviation from a linearity line of a D/A converter caused by the inability to express voltages in other than step functions

Quine-McCluskey algorithm: a method of minimization using tables instead of maps

race, logic: a problem introduced into a system by propagation delay through two possible paths to the destination

radix: the base of a number system; the number of symbols in the number system

RAM: random-access memory

ramp conversion: a method of A/D conversion by counting the time it takes for a reference voltage to equal the input voltage

random-access memory: a memory system in which all locations can be accessed directly, without having to access other locations first

rate multiplier: a divide-by-n counter where n is programmable

read only memory: a memory element capable of being read but incapable of being altered

record: a series of adjacent characters within a storage medium

reduction (of Boolean expressions): the process of determining the minimum expression capable of satisfying a logic problem

refresh: the act of periodically restoring degraded digital data to their original value

register: a memory element capable of containing one binary word

resolution: the voltage increment of a D/A or A/D converter that results from a change of one least significant bit

ring counter: a counter formed by circulating a 1 in a shift register whose serial output has been connected to its serial input

ripple counter: a counter formed by connecting each flip-flop output to the T lead of its next stage

ROM: read only memory

row dominance: a technique in tabular minimization of eliminating an implicant if it covers fewer of the same minterms than another implicant

RS flip-flop: a flip-flop with two inputs: a reset, which puts it in the 0 state, and a set, which puts it in the 1 state

RTL: resistor-transistor logic; a family of IC logic

RZ: return to zero; a digital magnetic recording technique

Schmitt trigger: a device with electronic hysteresis that is used for "squaring up" pulses

scratch-pad memory: fast-access temporary storage registers

self-complementing: the property of a BCD code whereby each binary code is the complement of its 9's complement decimal number

sense line: a line within a memory that senses whether the cell is a 1 or a 0

serial data: data in which bits are transmitted one after another over the same line

serial to parallel converter: a circuit that receives serial data and provides parallel outputs; usually a shift register

settling time: the time necessary for all logic elements to reach their final logic states

shift register: a series of flip-flops capable of shifting a binary number to the right or left

sign bit: the bit of a binary word that represents the arithmetic sign

single-shot: a monostable multivibrator

software: the programs and documentation of a system

SOP: sum of products

soroban: a Japanese abacus with one 5-bead and four 1-beads per digit

state: the binary number contained within a register

state diagram: a diagram which shows what binary numbers a register goes through

static memory: a memory whose contents remain indefinitely, requiring no refresh

strobe pulse: a pulse that samples binary information

successive approximation converter: a system of A/D conversion whereby the D/A voltage is generated by successive approximation

successive division: a process used for converting a decimal integer to another radix

successive multiplication: a process used for converting a decimal fraction to another radix

summer: an analog device used for adding currents or voltages

sum of products: a Boolean expression formed by **OR**ing **AND**ed terms

syllogism: a statement in formal logic used for arriving at a valid conclusion

synchronous logic: a logic system whereby all elements are synchronized to a master clock

tabular reduction: the Quine-McCluskey method of reducing a Boolean expression

terminal: a set of equipment that permits people to communicate with the computer

T flip-flop: a flip-flop element that toggles upon receipt of a pulse on the T input

time division multiplexing: a system of transmitting multiple channels over a single facility whereby each channel is assigned a recurring period of time

timer: a divide-by-n counter

toggle: to change a flip-flop to the opposite state

totem pole: an output of a gate that supplies current when low and receives current when high

track: one of the concentric circles formed by recorded data on a disk

transition: the edge of a pulse as it changes from one state to the other

tri-state logic: a type of gate output that can be either totem pole or high impedance, depending on a control lead

truth table: a table indicating the output of a combinational logic circuit for all input states

TTL: transistor-transistor logic; a family of digital integrated circuit logic

two's complement: a method of representing negative binary numbers in one's complement + 1 form

universal logic: any logic hardware that can be used as an **AND, OR,** and invert element

up counter: a counter that increments by one binary number each time a clock pulse is received

up/down counter: a counter capable of operating as an up counter or down counter, depending on a control lead

variable mapping: a technique of reduction in which variables are placed directly on the map

variables: input wires in a logic system

Venn diagram: a diagram that represents **ANDs** and **ORs** by intersecting circles

volatile memory: a memory whose contents are lost when power is removed

weighted code: a binary code in which each bit position is assigned a specified numerical value

word: a binary number

XS3: excess-3 code

ANSWERS TO ODD-NUMBERED PROBLEMS

CHAPTER ONE

1-1. Percent full scale is easily determined. It requires observer interpretation.

1-3. A method whereby information is expressed in measurable quantities.

1-5. The price of digital components continues to decline.

1-7. 1375

CHAPTER TWO

2-1. (a) 10	(b) 117	(c) 175
2-3. (a) 10,111	(b) 1,100,100	(c) 10,010,001
2-5. (a) 10,010	(b) 11,000	(c) 101,100
2-7. (a) 11	(b) 10,111	(c) 100,000
2-9. (a) 1,000,001	(b) 1,111,110	(c) 1,010,001,000
2-11. (a) 101 R10	(b) 111	
2-13. (a) 1011,1110	(b) 1011,1100,1001	(c) 1001,1011,1100,1000
2-15. (a) 15B7	(b) 2DDB	(c) FBAF
2-17. (a) 233	(b) 31,907	
2-19. (a) CE	(b) E23	
2-21. (a) E4	(b) 12,567	

2-23. (a) 1F (b) 74DA

2-25. (a) 5B82 (b) 348C

2-27. (a) 111,110,100,010 (b) 111,000,001,101 (c) 011,101,111,110

2-29. (a) 31 (b) 235 (c) 727

2-31. (a) 11,110,010 (b) 11,001,111 (c) 10,011,101

2-33. (a) 00,011,111 (b) 01,110,110

2-35. (a) 11,011.001,10 (b) 11.001,101,10 (c) 111,111.0110
(Rounded to nearest bit)

2-37. (a) 6.75 (b) 6.5625 (c) 21.3125

2-39. (a) 10,110,100,000 (b) 0.000,000,000,011,010,011

CHAPTER THREE

3-1. (a) 0011,0010,1000 (b) 0001,0100,1001,0111 (c) 1001,0111,0010,0101

3-3.   2421   5211
  (a) 0001,0100,1011   0001,0111,1000
  (b) 1101,0010,1100   1100,0011,1010
  (c) 1100,1101,0010,1011   1010,1100,0011,1000

3-5. (a) 5731 (b) 8629

3-7. (a) 3894 (b) 0271

3-9. (a) 0010 (b) 1001 (c) 1111

3-11. (a) 1101 (b) 100,110 (c) 110,101

3-13. 11,111,000

3-15. (a) 382 (b) 093 (c) 864

3-17. (a) 1,001,011 (b) 0,101,101 (c) 1,010,101

3-19. International: $-\cdots$ $\cdot\cdot$ $-\cdot$ $\cdot-$ $\cdot-\cdot$ $-\cdot-$ $-$
   American: $-\cdots$ $\cdot\cdot$ $-\cdot$ $\cdot-$ $\cdot$ $\cdot\cdot$ $\cdot\cdot$ $\cdot\cdot$

3-21. Teletypesetter

3-23. HERMAN HOLLERITH, BUFFALO NEW YORK

CHAPTER FOUR

4-1. Philosophy (rhetorical logic)

4-3. AND, OR, and NOT

4-5.

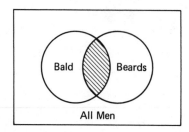

4-7. 8, 32, 1024

4-9. (a) 0   (b) 0   (c) 0

4-11. (a) W 0000 1111   (b) R 0000 1111
        X 0011 0011   S 0011 0011
        Y 0101 0101   T 0101 0101
        ───────────────   ───────────────
        T 0001 0011   V 0000 1110

4-13. Frozen   FFFF TTTT
        Leaks   FFTT FFTT
        Swim   FTFT FTFT
        ──────────────────
        Cross   TTFT TTTT

4-15. Shorts   FFTT
        Socks   FTFT
        ───────────────
        Trousers FFTT

4-17. (a) ABC (b) ABC (c) ABC (d) A (e) 0

4-19. (a) $A + B$ (b) B (c) BC (d) WX

4-21. (a) $A + B$ (b) 1   (c) $A\bar{B} + \bar{B}\bar{D}$ (d) $W\bar{X}$ (e) $AB + CD$

4-23. (a) A (b) $\bar{A}C + \bar{B}C$ (c) $\bar{A}\bar{B}$ (d) AB

4-25. (a)

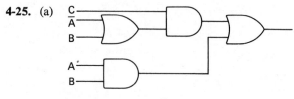

        (b)

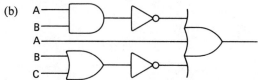

        (c)

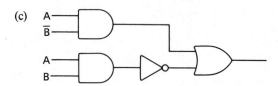

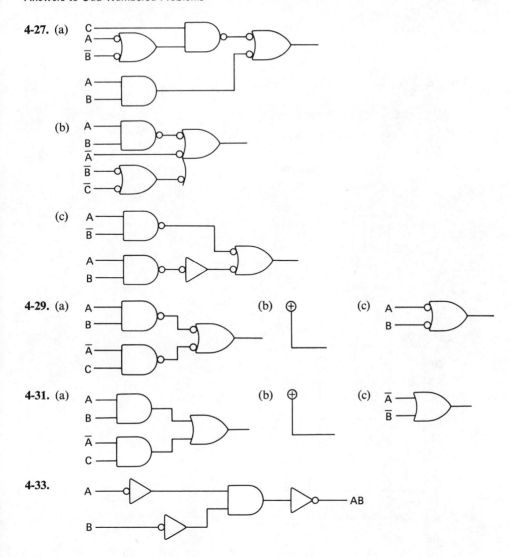

4-27. (a)

(b)

(c)

4-29. (a) (b) (c)

4-31. (a) (b) (c)

4-33.

CHAPTER FIVE

5-1. Voltage drop; leakage

5-3. Low input impedance, high power dissipation

5-5. Ten percent and ninety percent

5-7. Depletion and enhancement

5-9. 3.0 V; 0.8 V

5-11. RTL

5-13. TTL, CMOS

5-15. DIP

CHAPTER SIX

6-1. $\bar{A}\bar{B}C + \bar{A}B\bar{C} + \bar{A}BC + A\bar{B}C + ABC$

6-3. $\bar{A}\bar{B}\bar{C}D + \bar{A}B\bar{C}D + A\bar{B}\bar{C}D + A\bar{B}C\bar{D} + \bar{A}BCD + AB\bar{C}\bar{D} + AB\bar{C}D + ABC\bar{D}$
$+ ABCD$

6-5. (a) 5   (b) 10   (c) 25

6-7.

A\B	0	1
0	0	1
1	1	1

Truth Table Output: 0111

6-9.

AB\CD	00	01	11	10
00	0	0	0	0
01	0	0	1	0
11	0	0	0	0
10	0	1	0	1

Truth Table Output:
0000 0001 0110 0000

6-11.

$\bar{A}$

BC\DE	00	01	11	10
00	1	1	1	0
01	1	0	1	0
11	0	0	0	0
10	0	0	0	1

A

BC\DE	00	01	11	10
00	0	0	0	0
01	1	0	0	0
11	1	0	0	1
10	0	0	0	0

Truth Table Output:
1101 1001 0010 0000
0000 1000 0000 1010

6-13.

6-15.

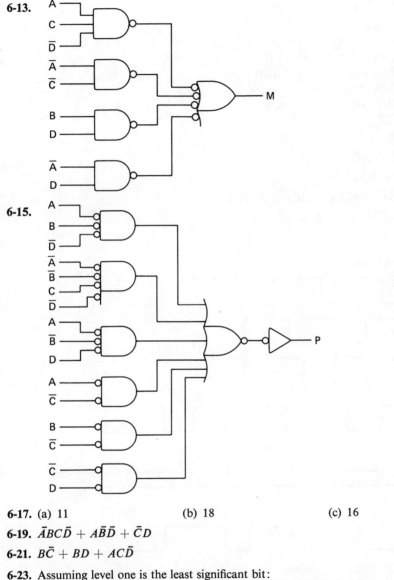

6-17. (a) 11 (b) 18 (c) 16

6-19. $\bar{A}BC\bar{D} + A\bar{B}\bar{D} + \bar{C}D$

6-21. $B\bar{C} + BD + AC\bar{D}$

6-23. Assuming level one is the least significant bit:
$\bar{A}\bar{B}CD\bar{E} + A\bar{C}\bar{E} + B\bar{C}D\bar{E} + BC\bar{D}\bar{E} + A\bar{B}C\bar{D} + \bar{A}\bar{B}\bar{C}E + \bar{B}\bar{C}DE* + A\bar{B}\bar{C}D*$

6-25. $R = (\bar{A} + B)(\bar{A} + D)(B + \bar{C} + D)$

6-27. $T = AC\bar{D} + \bar{A}BC + \bar{A}\bar{C}\bar{D} + \bar{A}\bar{B}D$
$T = (A + B + \bar{C} + D)(\bar{B} + C + \bar{D})(\bar{A} + C)(\bar{A} + \bar{D})$
POS form should be implemented.

6-29. SOP: $L = \bar{A}BD + ABC + A\bar{C}\bar{D} + A\bar{B}\bar{C}$
POS: $L = (\bar{A} + \bar{B} + C + \bar{D})(A + B)(A + D)(B + \bar{C})$

*Either term is correct.

6-31. $A\bar{D} + \bar{A}\bar{B}D + C$

6-33. $\bar{A}\bar{B}C\bar{D} + \bar{A}B\bar{C}\bar{D} + \bar{B}C\bar{E} + BCD + \bar{B}\bar{C}DE + A\bar{B}D$
$(A + B + \bar{C} + \bar{D} + E)(\bar{B} + \bar{C} + D)(\bar{B} + C + \bar{D})(B + C + D)(\bar{A} + \bar{B})$
$\cdot(\bar{A} + D + \bar{E})(A + B + C + E)*(A + C + \bar{D} + E)*$

6-35. $f_1 = A\bar{C}\bar{D} + ABC + AC\bar{D} + \bar{A}\bar{B}D$
$f_2 = A\bar{C}\bar{D} + B\bar{C}D + \bar{A}\bar{B}D + \bar{A}C$ or
$f_2 = A\bar{C}\bar{D} + AB\bar{C} + \bar{A}D + \bar{A}C\bar{D}$

6-37. $f_1 = \bar{A}\bar{D} + AC + AB$
$f_2 = A\bar{C}\bar{D} + \bar{A}C + \bar{A}D + AB$

6-39. $N = AB\bar{C} + \bar{A}\bar{B}D + \bar{B}C\bar{D}$

6-41. $Y = \bar{A}D + \bar{B}\bar{C} + A\bar{C}$

6-43. $Y = \bar{A}\bar{C}DE + B\bar{C}\bar{D}\bar{E} + ACF + AC\bar{D}$

6-45. $T = \bar{A}D + \bar{A}BC\bar{F} + BC\bar{D}G$

CHAPTER SEVEN

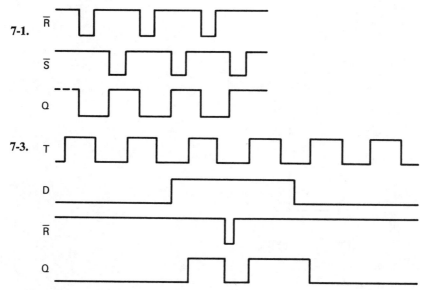

7-1. $\bar{R}$, $\bar{S}$, Q

7-3. T, D, $\bar{R}$, Q

7-5. Gate $= AB$; toggle off Q leads

7-7. $SET_A = SET_B = SET_C = \bar{A}\bar{B}$; toggle off $\bar{Q}$ leads

7-9. $JK_A = BC \qquad JK_B = C(\bar{A} + \bar{B}) \qquad JK_C = 1$

7-11. $D_A = B \qquad D_B = C \qquad D_C = \bar{A}\bar{B}\bar{C}$

7-13. $J_A = CD \qquad J_B = \bar{A}C \qquad J_C = \bar{B} + D \qquad J_D = B + AC$
$\quad\quad K_A = D \qquad K_B = C \qquad K_C = \bar{A} + D \qquad K_D = \bar{A}(\bar{B} + C)$

*Either term is correct.

7-15. $J_A = BD$ $J_B = CD$ $J_C = \bar{B}$ $J_D = 1$
  $K_A = BD$ $K_B = D$ $K_C = A + BD$ $K_D = \bar{A} + \bar{B}$

7-17. $J_A = BC$ $J_B = \bar{A}\bar{C}$ $J_C = A + B$
  $K_A = 1$ $K_B = C$ $K_C = 1$

7-19. $J_A = B(M + C)$ $J_B = M + C$ $J_C = \bar{M}$
  $K_A = B(M + C)$ $K_B = M + C$ $K_C = 1$

7-21. There are several possible designs. See page 376 for an example.

CHAPTER EIGHT

8-1. Tri-state or open collector

8-3. A Schmitt trigger

8-5. A one-of-eight decoder

8-7. An arithmetic and logic unit

8-9. Undefined

8-11. A retriggerable monostable multivibrator

8-13. 11 kHz.

CHAPTER NINE

9-1. (a) 64   (b) 65,536

9-3. A bus is a data path consisting of two or more discrete signal wires. It is a parallel path.

9-5. Only one chip is enabled. The data output for the disabled ICs are all set to high impedance.

9-7. The gate is surrounded by an insulator.

9-9. Using the three least significant bits of the ROM:

Loc.	Contents	Loc.	Contents	Loc.	Contents	Loc.	Contents
00	0000 0000	08	0000 0010	16	0000 0001	24	0000 0000
01	0000 0000	09	0000 0000	17	0000 0000	25	0000 0100
02	0000 0001	10	0000 0110	18	0000 0000	26	0000 0001
03	0000 0001	11	0000 0100	19	0000 0000	27	0000 0000
04	0000 0000	12	0000 0100	20	0000 0000	28	0000 0000
05	0000 0000	13	0000 0110	21	0000 0000	29	0000 0000
06	0000 0011	14	0000 0010	22	0000 0000	30	0000 0000
07	0000 0001	15	0000 0000	23	0000 0000	31	0000 0000

9-11. Bipolar – 2 resistors, 2 transistors (multiple emitters)
MOS RAM – 0 resistors, 6 transistors
MOS DRAM – 0 resistors, 1 transistor

9-13. (a) When $\overline{CAS}$ goes low.
(b) When $\overline{WR}$ goes low.

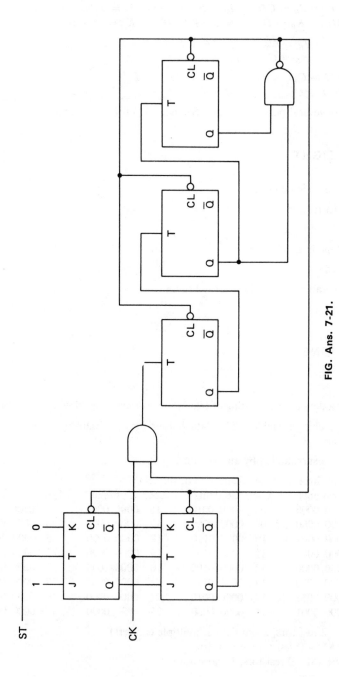

FIG. Ans. 7-21.

9-15. To accept the signal and to reject the noise

9-17. This permits these unused addresses to be used by other cards.

9-19. The processor will strobe the data into itself before it releases RD.

CHAPTER TEN

10-1. PC, IR, and RUN FF

10-3.

00 NBR1	10 JAZ 12	20 STORE 27
01 NBR2	11 JUMP 15	21 LOAD 08
02 NBR3	12 LOAD 26	22 IAC
03 NBR4	13 IAC	23 STORE 08
04 NBR5	14 STORE 26	24 JUMP 08
05 LOAD 25	15 LOAD 27	25 -5_{10}
06 STORE 27	16 IAC	26 0 (used as a counter)
07 STORE 26	17 JAZ 19	27 0 (counts iterations)
08 LOAD 00	18 JUMP 20	28 $00,101,001_2$
09 SUB 28	19 HALT	

10-5. A microprocessor unit contains all the basic functions of a computer except IO and memory.

10-7. Serial: requires fewer lines but is slower.
Parallel: requires more lines but is faster.

10-9. One of the IO lines could be used as a switch to select one of two RAMs.

10-11. Interrupts permit the processor to run without periodically checking for inputs. Further, interrupts permit high speed prioritization of IO requests.

10-13. The index register is two bytes wide and therefore requires two bytes of data, whereas the A register requires only one byte of data.

10-15. Faster access to any data on the media

10-17. Drum is fastest, tape is slowest.

CHAPTER ELEVEN

11-1. 15.4 V   20.4 V

11-3. 0.787%

11-5. 0.0488 V

11-7. 14 bits

11-9. The resistors must have weights of $2:1:2:4$

INDEX